# Lighting Historic Buildings

**Derek Phillips**

**Architectural Press**
An imprint of Butterworth-Heinemann

Architectural Press
An imprint of Butterworth-Heinemann
Linacre House, Jordan Hill, Oxford OX2 8DP
A division of Reed Educational and Professional Publishing Ltd

 A member of the Reed Elsevier plc group

OXFORD   BOSTON   JOHANNESBURG
MELBOURNE   NEW DELHI   SINGAPORE

First published 1997

**British Library Cataloguing in Publication Data**
Phillips, Derek, 1923–
    Lighting historic buildings: a prospectus
    1. Historic buildings – Lighting      2. Buildings – Lighting –
    History      3. Historic buildings – Remodelling for other use
    I. Title
    729.2'8

ISBN 0 7506 3342 5

Composition by Genesis Typesetting, Laser Quay, Rochester, Kent
Printed and bound in Great Britain by The Bath Press, Bath

# Contents

# Foreword

Lighting is an art form of the twentieth century, which grew in design practice from a basis of empirical theories that gave birth to knowledge. The sciences that now surround lighting shaped parts of how it was practised and by the middle of the century began to dominate most lighting installations in the western world. However, in the 1970s there was a so-called 'Energy Crisis' that caused a reexamination of all recommended lighting practices. The use of daylight was rediscovered and energy conservation became the guidepost of design. It is fair to say that energy conservation and care for the environment became universal concerns.

Now that we are reaching the end of the twentieth century, this book is a welcome addition to the literature on the subject. There were no inflated lighting standards when the work in this book took place. It brings into focus mankind's earliest use of daylight, the light of a flame and a vision of the illuminated spaces in those earlier times. It suggests ways in which to think about light in historic buildings and illustrates these ideas with wonderful examples and illustrations. It is also a great primer for finding ways to light well in an energy ethical manner today.

This book is for every educated person, not just architects, engineers, or historians, and is filled with information that can enrich your life's experience. It does so by providing examples of the light that people lived by in previous centuries. I strongly recommend that you mock up some similar spaces and accustom yourself to that environment. In many ways you will find that we have become an over-lighted society, victims to lighting habits we had no intention to create. Do so and you will become a participant in a noteworthy visual experience.

At the turn of this century, some books on lighting still contained technical sections on kerosene lamps, gaslight and electric arcs. Some books also had some very poetic aesthetic advice. For the most part that is all lost history and that is too bad. *Lighting Historic Buildings* brings that back by putting a sense of the light of everyday life of past centuries within our grasp. It then helps us to deal with it in our time and provides an instant reminder of how well we can see. That reminder can apply the brakes to the runaway lighting standards that evolved in the mid 1900s. I still find old public buildings, e.g. libraries, concert halls, museums, etc., where the lighting has not been changed, to be both comfortable to use and remarkably beautiful. They are beautiful because the lighting installed at the time was appropriate to the room and its use.

*Lighting Historic Buildings* is a pragmatic reference for the practising designer. It brings answers to questions or suggests means to achieve them and it is all done with basic empirical wisdom. For all the change wrought by the twentieth century there are still many people who have an appreciation and perhaps a devotion to the work done in previous centuries. There are still many places in the world that have changed very little and are now preserved by landmark preservation status. How fortunate. When change is required, however, the responsible alteration when made, should not intrude on what was previously built. Here, this book is worthwhile and fascinating reading.

Welcome to a reeducation on light and lighting.

Howard M. Brandston, Hon.,
FCIBSE, FIALD, FIES

# Acknowledgements

In the three years that it has taken to write this book I have received help from lighting consultants, architects, lighting equipment manufacturers and building owners.

It would be difficult to thank all those people who have assisted so I am restricting my acknowledgements to those kind people who have had the most influence on the book.

First of these is Derek Linstrum who read the book on behalf of the publisher and whose encyclopaedic knowledge of architectural history avoided much inaccuracy. Professor James Bell whose work in this field was an inspiration to me, helped both with text and photographs. David Loe from the Bartlett School of Environmental Studies ensured that the role of daylight was undiminished whilst my former colleague, John Howard, corrected the glossary.

Long discussions were held with designers Howard Brandston in the States, and Bill Allen whose work I much admire, and it would be churlish not to acknowledge Stanley Wells, whose book on period lighting has for many years been a mine of information.

# Introduction

The purpose of this book is first, to trace the history of daylighting and other sources of light up to the end of the nineteenth century to show the effect light has had on the form of buildings; and second, to establish how architects and designers today should approach the lighting of historic buildings in the context of present-day technology.

Buildings up to the end of the nineteenth century were dominated by the requirements of daylight. During the day, daylight was the only viable source while at night, light from the various sources available at the time was scarce, expensive, and often dangerous.

The world developed under daylight, and architects and builders learned to understand its geometry, to use its variability, and to create buildings which, first for functional and later for decorative purposes, were influenced in their plan and section by the introduction of natural light. Architects using natural light from windows achieved great sophistication. From the side, and later from overhead, windows were developed to provide the interiors of buildings with light having directional quality, sunlight and shadow while the window itself gave information of the world outside, the weather, the time of day and the season of the year.

The internal appearance of early buildings resulted from the daylight outside, acting upon the plan form and section of the structure, the building form being developed to modify the climate externally, and to introduce light to the interior. Tall interiors allowed tall openings, tall openings allowed light further towards the rear of a space. Low interiors reduced opening height, restricting light to the area close to the side walls. Overhead openings allowed light into the centre of the plan form, although having limitations upon its height.

Early artificial sources, consisting basically of different forms of fire – candles, oil lamps and firelight – had little effect upon the building form, with the exception of openings in the roof to let out smoke, when fire was used for cooking and heating. However, the decorative quality of artificial sources of light, associated with the manner of their support and containment, was discovered at an early stage. An interest was shown first in the candles and candelabra, lamps and chandeliers from which the light was derived, leading on to the design of wall or ceiling to which these might be fixed, the one reacting to the other. The whole development of interior design relates to a knowledge of and under-standing of the laws of light: the laws of reflection, transmission and refraction, its geometry and colour, and its capacity to provide information by means of light and shadow.

Daylight and the alternative means of 'fire' light, when thought of in terms of architectural form, remained entirely separate until the development of electric light, and the immense increase in the efficiency of modern light sources. The buildings considered as 'historic' demonstrate this separation. The later buildings of the twentieth century display a tendency to integrate, the form of the building being designed to use the functional qualities of natural light during the day, in association with those of electric light during both the day and night. Natural light and electric light are integrated to achieve a functional whole.

Historic buildings form a large part of the nation's heritage, whether used for their original purpose, as might be the case of a theatre, a railway terminus or house, where some modification may be required to meet its enhanced needs, while retaining the integrity of the original structure; or, alternatively, where the use has completely changed and it is reasonable to assume that the functional requirements of the new will outweigh the needs of the old, as might be the case of a redundant building transformed for an entirely different use such as an art gallery. There are many such conversions, and once the decision has been made to use the building in this way, it should still be possible to retain something of the original integrity of the building, without compromising the needs of the future.

This book's aim will have been achieved if it gives knowledge and understanding of the use of light in buildings of past periods, while providing a framework of how architects should approach such buildings today in the light of modern technology.

# 1 Daylight

There is no architecture without light and there can be no building where the presence of natural light, either in part or as a whole, will not benefit those who use it. It is impossible to overemphasize the important influence of natural light on the interior and exterior forms of buildings and on those who dwell in them. So daylight is the natural beginning.

Light in the earliest buildings was used by their designers in much the same way as it is today, for both function and decorative effect. In the Egyptian temple of Karnak, commenced in the third century BC, daylight was introduced at changes of roof level in the outer courts, with holes in the roof of the inner sanctums to provide shafts of sunlight onto statues, while ensuring that the incised surfaces of decorated walls reflected a more diffused light. Whatever the building form, whether pyramid, obelisk or temple, its external appearance resulted from the interaction of sun, sky and mass, where the mass was often designed for the specific purpose of reflection from gold, silver or light-coloured stone surfacing materials.

The 'rotunda' at the Pantheon in Rome, built by the Emperor Hadrian in the second century, is a circular domed building with a diameter of about 43 metres, having an 8 metre opening in the centre of the dome through which both light and weather can penetrate. As the earth moves around the sun, the sunlight pattern on the circular walls changes continually, giving a constantly varying appearance to the rotunda within. The oculus above would have had a symbolic meaning for the pagan gods for whom the building was erected in AD 120, but as an example of daylight used for both functional and decorative needs it survives today without peer. Externally the lower storey of the building was clad in white Pentelic marble with stucco above and the dome was originally clad in bronze, its circular form being modelled by the strongly directional sunlight.

Four centuries later the Christian church of Sta Sophia in Istanbul built by Justinian and considered to be the masterpiece of Byzantine architecture has a 32-metre diameter dome penetrated by a circle of 40 windows giving light to what at the time was a dome clad in pure gold. The contemporary Procopius said 'it was so singularly full of light and sunshine, you would declare that the place is not lighted by the sun from without, but that the rays are produced within itself. . .from the lightness of the building it does not appear to rest on a solid foundation, but to cover the place beneath as though it were suspended from heaven by the fabled golden chain'. Even allowing for the hyperbole of the time, it was clear that daylighting played a seminal role in creating the form of the building, providing functional light to the interior and, moreover,

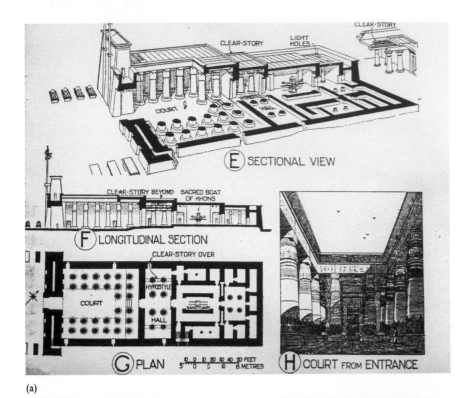

(a)

(b)

**Figure 1.1**
(a) The Temple at Karnak. (Copyright the Trustees of Sir Banister Fletcher). (b) The statue of Rameses II at Abu Simbel illuminated by a shaft of sunlight. (Photographer George Gerster, from *Light: from Aten to Laser* by Hess and Ashbery)

providing environmental lighting to the space by reflection from its decorative surfaces.

The innovations of structure in the Gothic period of the Middle Ages led to the wonder of the medieval cathedral, a structure of 'light', a magnificent tracery of stone between which natural light enters at all levels. The development of the flying buttress in particular permitted whole walls of window between, and the interiors of the cathedrals of northern Europe built between the twelfth and sixteenth centuries

**Figure 1.2**
The Pantheon, Rome. Interior, showing the daylight effect on its curved surface. (Copyright James Bell)

**Figure 1.3**
Sta Sophia, Istanbul. Interior of the building to show the daylight appearance. (Copyright Derek Phillips)

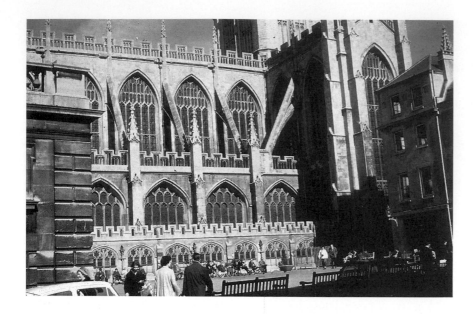

**Figure 1.4**
Bath Abbey illustrates the use of flying buttresses which allow large window areas. (Copyright James Bell)

epitomize the architecture of 'light', the tall interiors varying with the season, the time of day and the magic of sunlight.

Bath Abbey is just one example of a wealth of medieval ecclesiastical buildings which might be cited. It shows clearly the structure of flying buttresses, which allows whole walls of light to the interior.

But perhaps one of the most perfect relationships between daylighting and decorative surface can be found in the Baroque churches of southern Germany where the light from windows, concealed for the most part from the congregation, reflects from the white and gold surfaces of the interior, providing an almost magical appearance, like water cascading from a fountain. A fine example of this is the little church at Weiss, by Dominikus Zimmermann, in which the curvilinear plan expressed on the outside and inside of the church helps to explain the way in which the light is modelled in the interior. The church has fine ceilings by Tiepolo, which are generously lit by high-level windows.

So far, the examples quoted have largely dealt with ecclesiastical architecture, and while different building types will be explored later in the book (residences, public buildings, etc.) it should not be thought that daylight had little influence on other forms of construction. This is far from being the case – daylight influences the form of all building. The Italian palazzo is of interest since, while the facades give the impression of 'deep space', the plan form shows that the width of building was dictated by the need to introduce adequate daylight during the day for the functional needs of the inhabitants. Building widths of 15 metres with tall ceilings permitted bilateral daylighting, not dissimilar to some twentieth-century offices.

The effect of variation of climate is a major factor. In hot countries, the introduction of daylight is associated with the admission of heat. In the Italian palaces, the facade was often protected from heat by a two- or three-storey arcade, which both ameliorated the heat problem and softened the quality of light to the interior.

Summing up in the words of James Bell, 'Throughout history daylight has been a crucial factor in the design of buildings'. It is clear therefore that daylighting strategy is at the heart of building design, it is the starting point for buildings today as in the past.

(a)

(b)

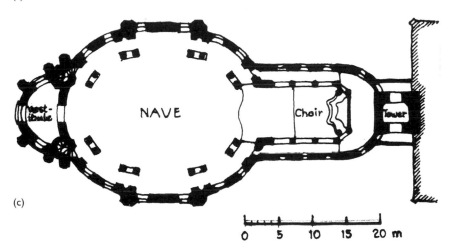

(c)

(d)

**Figure 1.5**
Die Weiss, Germany. (a) Exterior. (b) Interior.
(Copyright Derek Phillips.) (c) Plan. (Copyright
James Bell.) (d) Ceiling by Tiepolo

## ENVIRONMENTAL NEEDS?

The primary purpose of the window is to let in light to enable the purpose of a building to be fulfilled. But while the functional need for people to see well enough varies with the type, from a church to residence or factory, and can be evaluated in terms of performance, the environmental needs are less well defined, and for this reason are discussed first. They can be identified and listed as follows:

| | |
|---|---|
| Change | Orientation |
| Modelling | Sunlight |
| Colour | View |

### Change

This is the variation of colour and intensity due to the time and season of the year, together with its relationship with the environment beyond the building. The weather clearly has a large part to play, and summer sunshine and shadow, winter snow and clouds influence the way in

**Figure 1.8**
The Acropolis, Athens, seen under the actinic light of Greece. Emphasis is on the entasis on the columns and shadow patterns below the portico. (Copyright RIBA)

**Figure 1.9**
Overhead daylighting of Michelangelo's statue of David in Florence. (Copyright Derek Phillips)

## Modelling

While modelling can be considered an aspect both of orientation and of variation, since it derives from the directional and changing quality of the daylight, it deserves special emphasis. The capacity of daylight to give

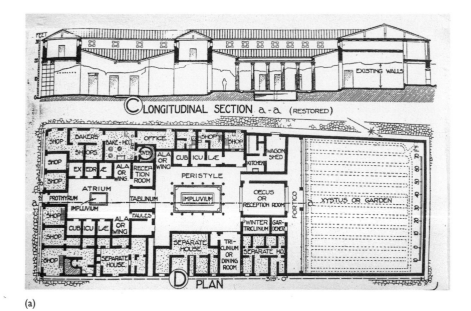

(b)

(c)

(a)

**Figure 1.10**
(a) Plan and section of a house in Pompeii. (b) Perspective drawing of the house of the Vettii at Pompeii. (Copyright the Trustees of Sir Banister Fletcher.) (c) Atrium with the central pool for catching water (impluvium) below the opening in the roof

form to an object by its directional flow gives that object its meaning, aiding our understanding. This might be the entasis on a Greek column given reality by the actinic light of Greece, a statue by Michelangelo lit from a roof light or the interior of a room created by the light from side windows. Its modelling comes from a combination of sun and sky outside.

## Sunlight

The introduction of sunlight to buildings in the western hemisphere is fundamentally good, but in warmer climates its thermal effects may need to be severely controlled. An example is the overhanging eaves and internal atria of the Roman villa where daylight is admitted but sunlight excluded. Compare this to its counterpart in western Europe, represented by merchants' houses in Amsterdam, where very large windows allowed maximum daylight. Sunlight, when present, is controlled by simple shutters.

Modification to sunlight entry may be required in certain building programmes where there are static workstations, and where at certain times of day sunlight can produce unacceptable thermal and visual conditions.

A knowledge of the sun path, related to the time of day and the date, will provide necessary information to a designer at the outset of the programme, and is of no less importance to the lighting consultant as to the architect. In the case of an historic building it allows studies to be made of the present situation, leading to modification where necessary.

**Figure 1.11**
(a) Windows in a house along a canal in Amsterdam. (Copyright Derek Phillips)

Building function will play a large part in determining the need for sunlight control, from the extreme of an art gallery where the artefacts on display must be protected from deterioration or fading, to a residence where sunlight is welcomed when it is scarce. In most situations the entry of sunlight is, however, to be encouraged both visually and therapeutically provided that the associated heat gain can be controlled.

Sunlight is a major contributor to all the factors already identified – variety or change, orientation and the modelling of interiors. Its influence on the view through the window is of further importance, and, seen from north-facing windows, the sun reflected back from a row of trees or building adds stimulation to the view. The nature of sunlight in different climates affects the exterior appearance of a building, and disappoint-

**Figure 1.11** (cont.)
(b) Interior of a bookshop. (Copyright *International Lighting Review*)

ment will result when a building designed for one climate is reproduced in another. An obvious example is a classical temple such as the Parthenon, detailed to react to the special climatic conditions of Greece, duplicated in the softer light of northern Europe. The exterior appearance will be very different. An awareness of climate is required to enable the designer to utilize the direction and intensity of light when determining the exterior form of a structure, and, more particularly, when detailing the openings for the admission of sun and skylight, 'the window'.

## Colour

Despite the fact that the 'physical' colour of objects and surfaces as measured by a spectrometer would be seen to vary from time of day and nature of sky, the colours of objects lit by daylight as perceived by the eye, are those which we believe. In a shop lit by electric sources a customer

**Figure 1.12**
Harrods Lace and Fancy Goods Department
(c. 1900); daylighting through second-floor void.
(Copyright Harrods Ltd)

will take a garment to the 'light', by which is meant daylight, to make a judgement as to its suitability. Daylight acts as the colour reference, since all other forms of light change the perceived colour in one way or another to a greater or lesser degree. Only daylight is thought to be the real colour, despite the fact that the colour of daylight itself varies from morning to night. Typical of this attitude was the development of the early department stores towards to the end of the nineteenth century, where daylight was encouraged to enter from overhead. An example can be seen in the early Harrods store in London, limited to two storeys, with voids cut in the upper floor allowing daylight to the lower. There is a return to this approach in large stores today.

## View

The 'view out' through a window, or how we perceive the world outside the building, is a dynamic experience associated with changes in daylight, sunlight and season. Any view is better than none, and it has been found that, in hospital, patients who have access to a view, however mundane, recover more quickly than those with no view.

At its lowest level, a view satisfies the physiological need for the eye to adapt and readapt to distance, while the content of the view is clearly of importance, reacting as it must to change. The view out assists people to orientate themselves. Blocking up, or the absence of view, is a design decision at the heart of daylighting strategy.

There are clearly some architectural programmes where it may be thought that a view out will lead to lack of concentration, such as in a school classroom, or perhaps a church. Here the windows may be planned at high level, satisfying the need for daylight to provide functional light, or may be concealed in order to introduce natural light to

**Figure 1.13**
Waddesdon Manor: the importance of 'view out' of the window. (Copyright Derek Phillips)

achieve special effects by reflection from the surfaces of the building, as in Baroque churches. 'View out' is not only important during the day, since if forms of exterior lighting are added beyond the window, the view can become important at night.

The detailing of the window, its size, shape and surrounds, are discussed later, since the comfort with which the view is experienced depends upon this. Indeed, in Japan the tendency to build a wall in front of a view, and then to make a hole in the wall from which the view is enjoyed, is the opposite to the 'picture window', where everything is revealed. A further aspect of view is the window as 'place'. The window would be less of a 'place' if there were no view out. An example of the bay window used as 'place' is that of the window seat at Château Chandon. In Palladian architecture, recessed windows would have a slab for 'sitting' placed on consoles either side of the window as a standard detail.

**Figure 1.14**
The window as 'place'. Window seat at Château Chandon. (Copyright Derek Phillips)

While 'view out' is a crucial aspect of daylighting, 'view in' should also be considered. During the day, except in special circumstances, the windows will appear dark, since the level of the light outside will be greater than that inside. At night this situation is reversed, and a view of what is inside will be revealed by the nature of the interior lighting. Different cultures tend to react in different ways, depending on the importance given to privacy. Examples of this are residences in Holland where at night curtains are often left open to display the richness of an interior with its indoor plants for all to enjoy. In Britain the tendency is to pull the curtains to shut out the night.

The preceding list of daylight criteria needs to be considered by the designer when setting out the daylight strategy for the building. There will, of course, be other environmental factors to be taken into account, in that the building form must react to the climate and external conditions of a country. But it is not the purpose of this book to examine the effects of heat, sound, pollution, etc. other than to acknowledge their importance.

## THE WINDOW: ITS FUNCTION

But what of the window itself throughout history? Encompassing any form of opening in the building fabric which lets in natural light, the window has taken many forms and the variety of window type is very wide. The question of security must have occupied the minds of the earliest builders, and splayed 'slit windows' had as much to do with the defence of the building as the entry of daylight.

Although there is evidence of glass making during the early civilizations of Egypt and Syria, this would not have been used for windows. Indeed, up to the fifteenth century in England, windows, where they were filled at all, mainly used other translucent materials, such as mica, horn, parchment, or oiled linen to fill the apertures. The development of glass in churches saw stained glass used from the twelfth century onwards to tell the Christian story, but the size of 'window quality glass' was quite small, with a maximum dimension of some 300 mm, and it was not before the late sixteenth century that windows became universal in domestic architecture, being formed from small panes in the 'leaded lights' we know today.

An interesting aside from the story of the development of the window is the attitude of government. In 1697 a window tax was imposed in England on all properties with more than six windows. This was in fact a tax on daylight, and since it did not stipulate the size of window, it tended to lead in the first place to blind or blocked-up windows, and, second, to larger windows. The tax lasted until 1851, and was a severe constraint on domestic building.

The window must be considered both from the point of view of the effect it has on the exterior appearance of a building – the building form – and the function served by it, associated with the interior appearance of the spaces it lights. The window as 'place' has already been mentioned, and the 'bay window' is a popular manifestation of this, often associated with a 'window seat'.

Climatic effect has a major part to play, and the small English thatched cottage with its minimal windows was a solution to climatic modification appropriate to the English climate and the affluence of the occupant, to be contrasted with the large windows of the great houses. The window clearly has a profound influence on the building's external appearance

**Figure 1.15**
Residential buildings, Malta, in bright sunlight. Minimal daylight openings show the influence of sunlight. (Copyright Derek Phillips)

from such extremes, on the one hand, as the medieval cathedral, where the structure itself has been designed to allow whole walls to admit maximum light, supported by flying buttresses and vaulting, often with lanterns above giving light to the central area; and, on the other, to the more solid exterior walls of buildings in warmer climates, where windows are designed to exclude heat while admitting minimum natural light.

It is generally true that throughout history, window size and disposition have been influenced as much by the climate of a country as by the pattern books of architectural fashion.

The question may well be asked as to what constitutes a 'well-daylit room'. This is perhaps the key question, but it is not answerable in purely engineering terms. Certainly all the environmental criteria already described play a part, the most important of which is that daylight is dynamic. The directional quality of daylight will be apparent even on a dull day, and the change in shadow patterns brought about by changes of the outside environment from time to time and season to season are to be welcomed. The following quotations may help to clarify what some people have thought to be good daylighting in an interior:

*A room appears pleasing to us because the play of light and colour evokes satisfying sensations* (*The Lighting of Buildings*, Hopkinson and Kay).

*A well-daylit room expands the boundaries of the space* (Howard Brandston, *lighting consultant*, USA).

*I can't define a space really as a space unless I have natural light. . .all spaces worthy of being called a space need natural light. Natural light gives mood to space by nuances of light in the time of day and the season of the year as it enters and modifies the space* (Louis Kahn, *architect*).

*For me it is one that provides a sensation of lightness, with a flow of light that is natural and gives me contact with the world outside* (David Loe, *Director of Lighting Studies*, The Bartlett, London University).

All the above quotations say slightly different things about daylight, all of which are relevant and add up to the impression that we have of a source of light which informs the space, gives a relationship with the outside world, is constantly varying, and achieves delight.

Having dealt briefly with the external appearance of the window, since this will become more apparent in later chapters dealing with specific building types, it is of importance to consider the different types of window, their detailing and the effect they have on the spaces they serve. The following types of window, each with its own characteristics, will assist in identifying both their main function and the interior quality associated with them:

Vertical windows
Horizontal windows/clerestorey windows
Window walls
Overhead windows
Concealed windows

**Figure 1.16**
The window at a medieval carpenter's shop, formed of overlapping small panes. (Picture from the Weald and Downland Museum, Singleton, Sussex)

**Figure 1.17**
Small window at Charlesworth Manor, showing Romanesque influence. (Copyright Derek Phillips)

**Figure 1.18**
Smallhythe Place, Kent: examples of horizontal windows in traditional construction. (Copyright Derek Phillips)

(a)

(b)

**Figure 1.19**
(a) Château de Chenonceau, France: exterior showing the bridge to the moat. (b) Château de Chenonceau: the interior of the curved embrasures to the 'bridge' windows. The embrasure balances the contrast of brightness between inside and outside, the windows articulating the view across the bridge. (Copyright Derek Phillips)

(a)

(b)

**Figure 1.20**
(a) Chiswick House, London, the Palladian villa modelled on the villa at Vicenza, Italy, by the architect William Kent for Lord Burlington in 1725. The windows around the dome give light to the Octagon Room. (b) Detail of window of Chiswick House, showing a taller central section (much copied in later buildings). (Copyright Derek Phillips)

**Figure 1.21**
A typical window from the 'loomshops' of the woollen industry in Lancashire, built at the turn of the nineteenth century. The horizontal seven-light windows gave excellent natural light. (Copyright W. John Smith FSA)

To introduce the subject of window design, a number of different styles of window from different periods is used to show the infinite variety available throughout history. While being very limited, they do illustrate some of the important features of the window.

## Vertical windows

These are the traditional window type, where it is desired to introduce light as far as possible into the space. Historic buildings tended to have higher interiors than is economically feasible today and for this reason daylight would reach deeper into the space. The nature and proportion of the window would be influenced by the contemporary structure and desirable plan form and available height.

The detailed configuration of the window and its surround was often splayed inwards to reduce the contrast between the brightness outside and the adjacent wall surfaces of the interior. One of the best examples of this was the Georgian window, where the interior splay was sometimes used to conceal shutters for security at night. The concept of splaying the wall dates back to early times, and may have had its origins in fortified buildings, particularly necessary where the walls were very thick. The splay allowed wider light distribution to the interior. Since large panes of glass were not available (indeed, early windows would have had no glass at all) the windows were broken up into small areas divided by glazing bars, these too being splayed. By splaying both the inside and outside of the 'bar' the apparent thickness of the window bars was reduced, the visual impression of the whole window being more open.

All the environmental qualities discussed can be satisfied by vertical windows. The view tends to be limited, or broken up into several consecutive sections. However, in a well-proportioned room there is little disadvantage in this, and such windows tend to provide a 'place' when related to furniture layouts. There will be a directional flow of light, its strength being dependent upon the nature of the day outside. It is this 'direction' which will inform the appearance of the surfaces and artefacts within the room and provide the characteristic appearance of a well-daylit room.

## Horizontal windows

This type of window tended to follow from structural advances, since it would have been impossible to achieve long runs of window where the external walls themselves were load bearing. Horizontal windows became more necessary where ceiling heights were reduced.

Typical of the horizontal window were those designed to provide functional daylight close to the external wall of early factory spaces. Provided windows were placed on either side of comparatively narrow buildings, sufficient daylight would have been available despite the lower ceiling heights. Other environmental advantages would also have been obtained, although on the south side sunlight might have been a problem and forms of blind introduced. In Figure 1.21, the cottages were built in 1820 and the front faces due south, with the rows of horizontal windows identifying the 'loomshops' at high level. Today the outer windows of the first floor are blocked to help thermal insulation.

(a)

(b)

**Figure 1.22**
(a) An Edwardian window at Crossrigg Hall seen externally. (b) The interior lit by the window at Crossrigg Hall. (Copyright Jo and Fred Wills)

The interior of spaces adopting horizontal windows is dependent upon the height of the windows, and rows of smaller windows or 'clerestoreys' at high level were developed to smooth the more directional quality of the windows at low level by lighting the ceilings or roof, and by inter-reflection, giving a more gentle overall light to the room, with a deeper penetration of light. They also prevented a 'view out' where this was considered undesirable.

## The window wall

An earlier structural innovation made possible by the removal of the wall and the introduction of flying buttresses in the English cathedral is where the whole wall became the window, and the interior was suffused with light. This light was often modified by stained glass telling the Christian story and adding colour to the interior.

Walls of light were also introduced into residential buildings by means of crosswall construction, eliminating the need for 'end' walls, which

**Figure 1.23**
Kedleston Hall, Derbyshire. The saloon lit by the daylit dome. (Copyright National Trust Photo Library)

were then filled with glass. Modern glass technology has greatly improved the capacity of the window wall in reflecting or otherwise modifying the external climate. But such sophistication would have been unavailable to the designers of historic buildings, and there is a danger that if such techniques are substituted today, the appearance of the building may be adversely affected.

## Overhead windows

Light from above was not new in that in the early Egyptian temples the gaps between roofs at different heights permitted some light to penetrate the exterior courts of these buildings. In the great thermae or public baths in Rome, such as those of Diocletian or Caracalla, top light was obtained by clerestorey lighting, adding to the lofty and impressive interiors. But it was in the seventeenth and eighteenth centuries that the provision of light from above, by means of laylights or glass domes, freed up the plan,

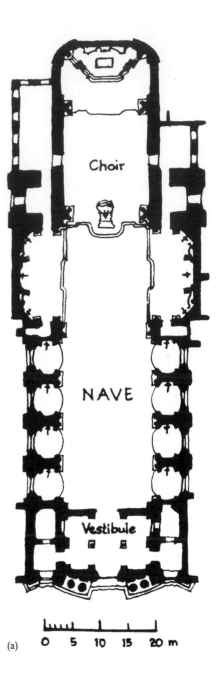

(a)

(b)

**Figure 1.24**
Zwiefalten Abbey. (a) The plan illustrating the louvre effect of the windows to the nave, concealing the daylight openings. (b) Interior showing the daylight effect. (Copyright James Bell)

allowing deeper buildings. Developments in structure in the nineteenth century allowed the large glazed roof coverings to railway terminals, industrial buildings, or Paxton's Crystal Palace or to give light to the interiors of the large department stores of the period.

## Concealed windows

This is not so much a category of window type as of the manner in which the window is used. Here our concern is with the indirect effect of the window, whether vertical, horizontal or overhead.

Concealed windows can be used in a strictly functional manner, for the lighting of works in an art gallery or for atmospheric effects in a Baroque church, where the surfaces of the interior are bathed in light from windows concealed from the view of the congregation. By its nature the concealed window denies the advantage of a view, and while certain of the environmental factors have less impact, they are all present to a degree, particularly 'change'.

The categories illustrated above are very wide and can be broken down into many subcategories, but they are of help when studying the factors which make up the components of daylight design.

## DAYLIGHT DESIGN

In the same way that a lack of a scientific approach to structural design did not stop the building of the Pantheon or the Pont du Gard, so the lack of such an approach to daylight calculation did not impede the progress of very sophisticated daylight solutions. There is little evidence of a scientific approach to daylighting until the twentieth century, but looking at the daylighting of historic buildings such as Sta Sophia in Istanbul it is difficult to imagine that this could have been achieved without a significant practical knowledge of the subject.

The structures passed down to us throughout history are those that survived, and there is little doubt that many would have failed. So it must have been with daylighting. The structural analogy is valid, since the incorporation of openings or windows in the walls is implicit in the form of the building.

The design of daylighting would have been based on observation, on trial and error, a form of experimentation upon which an empirical theory would have been built up. The empirical judgements made would no doubt have changed, from the needs in early buildings for protection and a view of external danger to the more sophisticated requirements to create satisfying interiors.

One of the design tools used by students of daylighting today is the use of simple models, since the relationship of the external light to that available internally is the same at whatever scale, whether model or full size. By carrying the model outside it is possible to judge the internal appearance of a space in different states of weather, times of day, and at different orientations. It seems inconceivable that when an important building was planned, experiments upon these simple lines were not carried out.

Therefore while a scientific approach to daylighting design did not emerge until the 1930s, 'daylighting design' was alive and well at a much earlier date.

# 2 Light sources other than daylight

While daylight results from the sun and the sky, the sources of 'artificial' light, up to the end of the nineteenth century, derived from 'fire'. Objection can be made to the term 'artificial' when applied to fire, since fire is one of the elements, but the term is well enough understood to refer to those forms of light which are used in the absence of daylight, and I shall use the term 'artificial' in this context.

The first uses of fire were for heating, cooking and light, although initially the use of fire for light was no doubt subsidiary to the first two. Primitive people would have tended to get up with the dawn and go to rest at sunset, until they discovered that torches made from branches dipped in animal fat could extend their day.

Early oil lamps, such as those found at Lascaux in France, are over 15 000 years old and would have been used by the cave painters of the period. These lamps, of hollowed-out stone, clay or shells filled with fish or other oil, were the precursors of the more sophisticated oil lamps which, together with candles, were the principal form of artificial light up to the nineteenth century.

It has been said that the use of a wick set in oil, burning totally and exclusively for the purpose of giving light, was as revolutionary in the development of lighting as the wheel was to the history of transport. There are excellent guides available detailing the infinite variety of lamps developed, but to catalogue these is not the purpose of this book, other than in the broadest sense (see *Period Lighting* by Stanley Wells). It is the extent to which their light affected the appearance of the spaces they served, rather than the appearance of the lamp itself, which is our chief concern. This is not to deny the importance of the vehicle for the light, but this is secondary to its impact on the space.

Very broadly the types of lamp were as follows:

1 Oil lamps: ranging from the earliest types relying on oil with a form of floating wick to the sophisticated lamps of the eighteenth or nineteenth centuries with wicks designed to increase the amount of light.
2 Candles: developed from rush lights, were common from the sixth century. They ranged from tallow moulds derived from animal fat to the more expensive candles developed for church use and used in the houses of the rich.

3 Gas lamps: available from the beginning of the nineteenth century. These were the greatest achievement during the period up to the age of electricity, disassociating the source of energy from the light itself.
4 Electric lamps: available towards the end of the nineteenth century, these fall just within the period of 'historic buildings' covered by this book, but cannot be described by the term 'flame' source, despite being initially a 'heat' source.

## OIL LAMPS

To separate the primitive oil lamps from the more sophisticated argand type developed in the eighteenth century is to distinguish between a low level of light, as from a candle, to the much brighter light obtainable from later inventions. Pan lamps, consisting of hollowed-out stone or clay, or a shell containing whatever oil was available, were used since fire was discovered, and float lamps of many descriptions followed, in which much ingenuity was displayed in forms of decoration. However, apart from the method of support and the number of wicks used, they provide an identical quality of light.

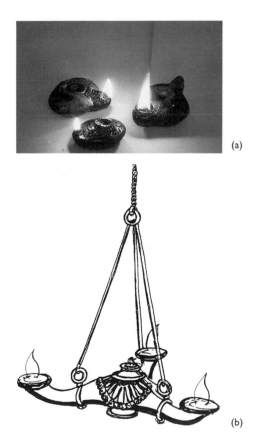

(a)

(b)

**Figure 2.1**
Early light sources (a) Oil lamps and float lamps, the most primitive early sources of light, used animal fat or fish oil with some form of wick such as moss. (b) Roman Pendant, based on sketch by Stanley Wells. (c) Dutch oil lamps showing similarity to the Roman Pendant. (Copyright Science Museum, London.)

(c)

Some lamps were portable, and could be carried around a property to negotiate ill-lit corridors, halls or staircases while others were wall mounted or on some form of pedestal from the floor, rather like a present-day floor standard. Others such as the Roman lamp from Pompeii of the first century were pendant; this type of lamp might have a central oil font, feeding a number of arms. The only way to increase the amount of light was to increase the number of wicks.

The magic of the early oil lamp was that it gave light, albeit very little for functional purposes, and it was only when large numbers of these were used that they would have contributed to the interior appearance of buildings in any decorative sense, despite the fact that individually many of the lamps were beautifully crafted. However,

**Figure 2.1** (cont.)
(d) Sixteenth-century kitchen, with hanging oil lamps. (Line drawing from *Flickering Flame* by Leroy Thwing)

from the lighting point of view, the appearance of a room would not greatly alter, whether the light derived from a simple pan lamp or one of ornamented silver. By day natural light would have been the main illuminant and informant of the spaces within the building. At night there would have been pools of light wherever an oil lamp was placed, but while providing light for safety it would have given scant information of the spaces lit.

As the light from an individual lamp would have been minimal, multiple lamps were designed to obtain the higher levels of light considered to be desirable. The menorah, the Jewish religious lamp with its seven arms derived from instructions in the Bible, would have used oil lamps originally, later converted to candles, the additional arms producing a higher level of light for the ritual.

It was not until towards the end of the eighteenth century that we see the invention of the argand lamp; which, by using the new theories of combustion developed by Antoine Lavoisier in 1770, allowed new technology to improve the light output of the oil lamp, and to enable it to be more easily controlled. The considerable improvement in the design of oil lamps themselves meant that they could begin to have some effect upon the decoration of the rooms in

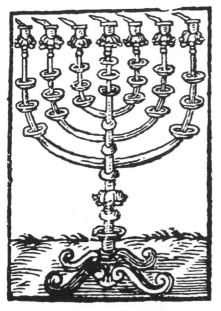

THE CANDLESTICKE.

(e)

(g)

(f)

**Figure 2.1** (cont.)
(e) The menorah, or Jewish candlestick, derived from information in the Bible. Traditionally, this has seven arms. (f) The argand lamp. (g) The 'student' lamp (1830), a model for many lamps that followed up to the present day. (Copyright Science Museum, London)

which they were placed. However, their influence was felt more in terms of their function, allowing a greater use of space after dark, rather than contributing to its form or decoration. The later introduction by 1869 of paraffin (kerosene) as the oil used made a safer and brighter illuminant.

The so-called 'student lamp' was introduced from France in 1830, and became very popular. It employed Colza oil, contained in the top metal flask, fed by gravity to the wick. This type of lamp has been copied right up to the present day and with its wide glass shade gives a pleasant and portable light.

An interesting comment on the social importance of lighting in the nineteenth century is indicated by a Parisian 'immeuble' or section through a building showing how the amount of light was strictly related to the hierarchy of a household. The greater the importance of the room, the more light, leaving the utility areas or servants' quarters in semi-darkness.

## CANDLES

Candles have a place in the history of lighting from earliest times. There was no linear development with each country proceeding at the same pace, so it is necessary to choose developments from different countries to establish the overall trends.

The earliest form of candle was the rushlight or taper, formed by splitting a rush and dipping it once or twice in some form of melted fat, until hardened when it could be lit. It gave about the same amount of light as an early oil lamp, but would have tended to flicker and burn for a short time only. It is thought that these were available as early as 600 BC, and illustrations from Sweden of people holding these rush lights or 'splints' in their mouths to light their work are available as late as the fifteenth century. More generally the rushlights would have been held in metal 'splint holders'. These were strictly functional, unlike the decorative candle holders which were to follow.

The tallow candle dates back at least to AD 500, when wicks of different material were dipped in tallow or formed in 'moulds', a name by which the tallow candle was known. These had the advantage of burning for longer hours, but being made of tallow, a product of beef or mutton fat, they would have tended to be smelly and smoky, they guttered, and were difficult to maintain alight. Tallow candles were the poor people's light, and despite the development of the candles we know today, they persisted to such an extent that the Tallow Chandlers Company was formed in London in 1463, and became one of the great city livery companies.

The church was the principal instigator of the wax candle, requiring candles which would burn for many hours, so that large candles similar to those we see in some churches today became available. These wax candles were expensive and were otherwise seen only in the houses of the rich.

The best form of material for candle making was spermaceti, taken from the head cavity of a whale, and a candle made from this was used to measure the intensity of light, from which derived the unit of light or 'candle power'. There was little difference in the amount of

(a)

(b)

**Figure 2.2**
(a) Candles. Early use of 'splints' in fifteenth-century Sweden (line drawing from an early print). (b) Sixteenth-century German stained glass picture, illustrating a bedroom of the period with a single candle. (Copyright Victoria and Albert Museum, London.) See also Bayleaf House (4.2f)

(c)

(d)

**Figure 2.2** (cont.)
(c) Brass candelabra (1435) similar to that in the painting 'The Annunciation' by Rogier van der Weyden in the Louvre, Paris. (d) The Gloucester Candlestick dating from 1110. (By courtesy of the Board of Trustees of the Victoria and Albert Museum, London)

light given from a single candle of whatever material, or even an oil lamp, from prehistoric times, remaining virtually the same until the growth of technology in the eighteenth century. There is little need therefore to differentiate between the materials from which candles were made. The development of candle light can be taken as a whole.

Early candle holders were simple functional objects, and even when multiple candle holders were developed to give more light, these were often in the form of suspended wheels with a number of holders mounted along their circumference. Whether the holders were 'prickets', or spikes to which the candle is fixed, or circular holders into which the candle is placed, the lit effect was the same.

Until the Tudor period in England, the early sixteenth century, the poor would still have been using oil lamps or rushlights, but by this time ornate candlesticks, pendants and wall sconces were being designed for candles. A wonderful example of this is the Gloucester Candlestick designed for the church as early as the twelfth Century.

Despite the infinite variety of ingenuity in creating all manner of designs for candleholders they reduce to three main types: the pendant, the wall mount, and the portable floor or table mounting. All had one element in common. They were designed to be at a level where they could be reached to enable the candles to be maintained and replaced safely.

## THE PENDANT

This is the most versatile of the three, where the number of lamps dictated the amount of light available. The candles were sometimes placed on arms originating from a central boss, or related to the crystal drops of glass chandeliers only limited by the invention displayed by the craftsmen and designers of the day. Many different materials were used in their manufacture, from simple wooden wheels to beautifully decorated bronze silver and ormolu pendants. The height of the pendant was limited to that which could be maintained from a ladder, although in some cases the pendant would be suspended by a pulley to enable it to be raised out of reach, and lowered to a more manageable height.

There are two aspects to candlelight. The first is the effect it has on the building, or the lit appearance of the space the light creates. The second is the extent to which the method of support, the candle holder, candelabra, or chandelier, provides a part of the decoration of a space, in the same way as the ceiling, walls or furniture. This may be insignificant, but where the effect has been considered it can be overwhelming. The two aspects must be taken together, since one will inform the other to a greater or lesser degree.

As an example, chandeliers provided not only light, but the light which was reflected from the cut glass or polished metal surfaces gave a specific quality to the chandelier and to the space in which it was hung. Chandeliers or wall sconces became an important part of the decoration of the room, and in the carefully designed interiors of the eighteenth century the placing of the lights was integrated with the interior design, with the wall panelling, and ceilings, as might be seen in the interior designs of the Adam brothers.

The reconstruction of a fifthteenth-century Renaissance party in a recent Italian film gave a dramatic impression of what such a room would have been like lit by pendant chandeliers...the smell and heat, wax dripping on to the heads of the participants, the pools of light below each pendant, with more light up to the ceiling than down. Then when the party is over, the servants lower the chandeliers to snuff out the candles.

Such a scene must have been one of magic, with the shifting shadows, the colour and garments of the participants reflected from the overhead candlelight, and despite some discomfort it must have been most exhilarating. A world apart from the experience of the poor of the period with perhaps a single tallow mould or oil lamp to be used by the whole family after nightfall.

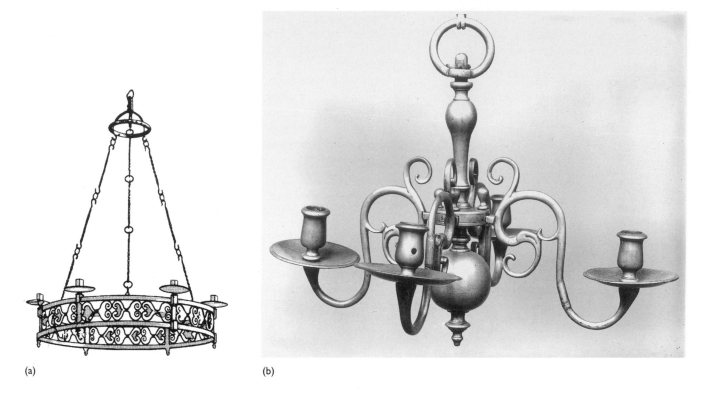

(a)

(b)

(c)

**Figure 2.3**
Pendants. (a) A Tudor 'corona' pendant with candle holders. (Adapted from *Period Lighting*, by Stanley Wells.) (b) A Dutch pendant with candle holders. (Copyright Victoria and Albert Museum, London.) (c) A French crystal chandelier, made from Waterford crystal and installed at Kedleston Hall, Derbyshire. (Copyright National Trust)

**Figure 2.4**
Wall and floor mounts. (a) Silver wall sconce,
Temple Newsum. (Copyright Derek Phillips)

## THE WALL MOUNT

Very simple wood or wrought iron brackets were used to mount candles initially, rather similar to the metal squints used for rushlights. But after a time these became more decorative, more arms were added, and one of the most important developments was in the addition of reflective backs, in silver or mirrored surfaces, to reflect back the light which would otherwise have been wasted. These 'wall sconces' became a part of the rich decoration of a period, improving the light output, but more than this, were related to the panelling details of the wall surfaces which they adorned, thus becoming an important aspect of the interior design.

## THE PORTABLE FLOOR OR TABLE MOUNT

These varied from the simple hand-held candle holder, which could be carried by a person around a house, to more ornate floor- or table-mounted lamps often with a number of arms. The latter gave larger pools of light and might have been used in dining rooms at meal times and then moved to other parts of the house as the need arose. Table-mounted lamps were often used on chimney pieces, or fixed to the newel posts of staircases in the more affluent houses.

The amount of light directed on a person's work, such as lace making, could be enhanced by means of glass globes or 'condensers'. These were used by lace makers after dark to make a single candle give light for two or three workers.

The seventeenth century appears to have been one of 'conspicuous consumption', and in 1608 the park at Versailles was lit up by 24 000 candles, an unimaginable number, and one has to ask how they were kept alight, or changed as the night wore on. To summarize the use of candles; it is true to say that from the earliest candles to sophisticated candelabra, the candle gave an identical light for rich and poor alike, but the rich had the advantage of having more of them, and so more light.

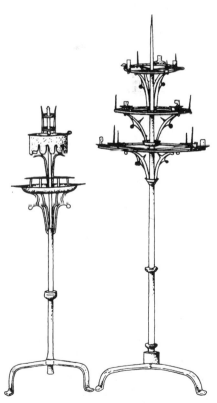

**Figure 2.4** (cont.)
(b) Fifteenth-century floor mount with prickets, in simple wrought iron (adapted from *Period Lighting* by Stanley Wells)

**Figure 2.4** (cont.)
(c) Candle condenser fitting used in lace making. (Copyright Science Museum, London)

**Figure 2.5**
'Masquerade in the Haymarket' (painting by Giuseppe Grisoni (1669–1769)). It was said that 500 wax candles were employed. (Copyright Victoria and Albert Museum, London)

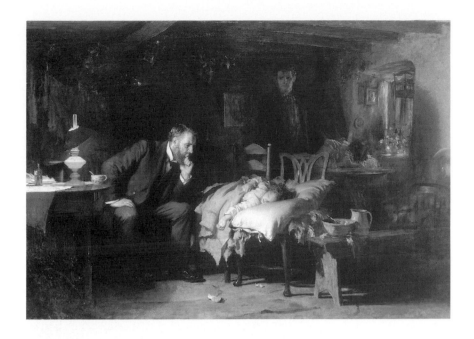

**Figure 2.6**
'The Doctor', painted by Sir Luke Fields in 1891.
This is good evidence of the use of a single oil
lamp for a doctor attending a sick child (poor
households would have used a single candle).
(Copyright The Tate Gallery, London)

## GASLIGHT

While experiments had taken place to produce light from burning coal
gas in the early years of the eighteenth century it was not until the
needs of the Industrial Revolution (around 1800) that the commercial
use of gaslighting became a reality. By 1805 gaslight was used by the
owners of cotton mills, previously lit by oil lamps. It was the
imperative of lighting up the mills and workshops which brought
ingenuity to bear on the production of a cheaper and more reliable
source of light than was at that time available in the candles and oil
lamps of the day.

**Figure 2.7**
Gaslighting. (a) An example of early gaslighting in
a factory. (Copyright Sugg Lighting)

**Figure 2.7** (cont.)
(b) Bradford railway station lit by gas. (Copyright Sugg Lighting)

The engineer, William Murdock, speaking of gaslighting in 1802, said:

> *The peculiar softness and clearness of the light brought it into great favour with the work people, and it being free from the inconvenience of sparks, and the frequent necessity of snuffing of candles . . . as tending to diminish the hazard of fire.*

Other advantages were said to be its reliability, its uniformity, and, in the industrial context, its lower cost and insurance premiums. It was also capable of control so that less or more light could be obtained. Despite

**Figure 2.7** (cont.)
(c) Gaslighting in an office. (Copyright Sugg Lighting)

**Figure 2.7** (cont.)
(d) Tooting Market, London, lit by gas. (Copyright Sugg Lighting)

these advantages, and because gaslight had been seen initially as applying to industry there was some reaction to it when it came to lighting in houses. It was thought to be socially inferior, and even when used in residences, tended initially to be restricted to the domestic quarters.

The advantages of separating the means of combustion – the gas – from the light itself ultimately became overwhelming, and once a stable gas supply was available, by means of the installation in the cities of gasometers, gas became universal by the middle of the nineteenth century. Prince Albert had gaslighting installed at Windsor in 1850 and despite reservations as to its heat and its injurious effect upon decorations expressed as late as the 1880s, candles became reserved, as they still are, for special occasions where their qualities of sparkle, warmth and flicker make them uniquely acceptable where levels of light can be low, as for dining.

Exterior lighting using gas became readily accepted, and when gas light was installed by the Prince Regent at the Brighton Pavilion, it was first used to provide light outside the stained glass windows, with a 'magic' light shining into the interior, giving the impression that the spaces were lit from inside. It was not until after the Pavilion was sold in 1850 to the Brighton Corporation that the chandeliers were converted to gas, later in 1883, being converted to electricity to avoid the deterioration of the internal fabrics.

The Mall in London was lit by gaslight as early as 1807, only to be overtaken by electric light half a century later. From 1809 onwards there was a rapid growth in gas supply, and by 1823, 215 miles of London streets had been lit by gas, using some 40 000 gas lamps. Four supply companies with a capacity of a million cubic feet of gas had been established in London, with other cities creating similar facilities. As supplies multiplied, with local gasworks installed in smaller towns, so did the use of gas reach into new fields. Gaslighting within buildings became the norm, and was even used for floodlighting. However, our concern is principally the use of gaslight in buildings, and here there were no great changes in the methods employed. More light was produced,

**Figure 2.7** (cont.)
(e) Gaslighting in a church. (Copyright Sugg Lighting)

# 3  Analysis of the lighting problem

The original architect will have conceived the building as lit by natural light during the day, with the addition of some form of flame source – oil lamps, candles, or gaslight – at night. It would not have occurred to the architect to design artificial light sources to combine with daylighting to achieve an integrated approach as is possible today, nor would it have seemed necessary.

The little cathedral at Iona is an example of the separation between daylight and flame sources, illustrating the appearance during the day with light reflected from the warm stonework of the church, providing adequate light for the congregation. At night the whole atmosphere changes with a concentration of light from candles at the altar, as it would have been, providing a sense of mystery and awe after dark, concentrating on the ritual of the church. In the interest of safety the candles are now electric and lack the sparkle of the candles in use earlier.

First and foremost historic buildings were 'daylit', natural light being encouraged to enter the building from the outside by means of different types of opening. It must be our first concern when dealing with the problem of what to do with the lighting to make a study of the windows, laylights and clerestoreys related to the form of the building, to analyse their impact, before considering what needs to be done (if anything) to provide other forms of light both during the day and at night.

Daylight, and particularly sunlight, while being wholly welcome, do have their obverse sides. In addition to the environmental problems of heat gain and noise, there are glare and the deterioration of degradable substances due to ultraviolet light, which may need addressing. For reasons of conservation, means of control may have to be incorporated into museums and art galleries – indeed, in any historical building, housing sensitive materials. Dutch artists like Rembrandt controlled the light on the subjects of their paintings by manipulating the various moveable shutters available at the time.

So before the lighting designer considers methods for the night-time lighting of the building, he or she must first be aware of the light of day as it reflects off the interior surfaces, giving the characteristic flow of light which results in its daylit appearance. This is not to say that the night appearance must strive to be the same, far from it. Night has its own special atmosphere. Rather, the qualities we associate with daylight

(a)

(b)

**Figure 3.1**
(a) Iona Cathedral, no larger than a parish church, seen under daylight. (b) The same view at night, showing the altar and choir stalls lit by candlelight, a total separation of the day and night effects. (Copyright Iona Community)

should be understood before the designer sets out to create the appropriate night appearance.

## THE NATURE OF THE PROBLEM

There are three different lighting problems in historic buildings related to their use.

1 When the building is to be retained in its original form, and enjoyed by its occupants, or perhaps by visitors for its historic value
2 When the building is to be retained in its original form but used more extensively to meet present-day needs, requiring modifications to the lighting design

3 When the building is designed for an entirely new use. There is a growing need for this type of solution, and it often presents the greatest challenge.

## The building retained in its original form

This is perhaps the simplest problem, for whether the building is used exclusively by its present occupants or opened to visitors, the lighting needs will have changed little since its inception. The primary lighting will be from daylight during the day, and where the daylight was adequate for its original purpose, whether an historic house, castle or palace, court house or church, then during daylight hours there will generally be sufficient light to enjoy the spaces as visualized by the original designer or architect, although there may be parts of the building that require additional light for safety. The interior appearance of the spaces within the building will change with the seasons, the time and type of day. It is this variety which marks one of the chief characteristics of daylight.

Where the building is used after dark, as in the majority of cases, then the lighting appropriate to its period should be adopted and if this results in a comparatively low level of light derived from the chandeliers, wall brackets and portable lamps, then this may be all that is desirable, albeit for reasons of safety the original flame sources may be electrified. At night a person's adaptation level is lowered and satisfactory results can be obtained from quite low levels of light.

What is important is that the appearance of the spaces should represent as far as possible the original aesthetic experience. It is of particular importance that whatever hardware is used for night lighting does not destroy the daylit appearance. In residences there was often so little light available, with lamps or candles being carried from room to room, that a completely authentic experience

**Figure 3.2**
The courthouse at Long Crendon by daylight, which is wholly adequate. No attempt is made to light it at night and visitors are denied access. (Copyright Derek Phillips)

(a)

(b)

(c)

**Figure 3.3**
The Court Hall at East Meon, built by William Wynford for William of Wykeham, Bishop of Winchester (1367–1404). (a) An exterior view of the Hall to show the tall windows, which were not glazed for some years after the Hall was built. (b) The interior of the Hall by daylight, which is very adequate. At night additional electric sources are brought in. The 'corona' candle fitting was added at a later date. (c) The 'solar', now used as a library by the family and as a meeting room. Lighting has been concealed by the timber roof structure. (Copyright Derek Phillips)

would be impossible, if the building were now occupied or open to the public. There may be a case for allowing only limited access, to small parties, where light from flame sources is carried around from room to room, as is done sometimes on the Continent.

The courthouse at Long Crendon, Buckinghamshire, is an example of a fourteenth-century building no longer in use and where what flame sources there may have been have been removed. The building is open to the public during daylight only.

The Court Hall at East Meon, Hampshire, is an example of a fourteenth-century building which is now privately owned and used by the family. The hall itself was one of the manors of the Bishop of Winchester, and when built, there was no glass in the large windows. The Court Hall is used occasionally by the local community, with additional lighting added, for musical recitals. The house associated with the Hall is occupied and concealed artificial lighting has been added as, for example, in the library.

(a)

(b)

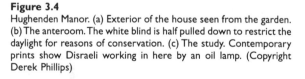

(c)

**Figure 3.4**
Hughenden Manor. (a) Exterior of the house seen from the garden.
(b) The anteroom. The white blind is half pulled down to restrict the
daylight for reasons of conservation. (c) The study. Contemporary
prints show Disraeli working in here by an oil lamp. (Copyright
Derek Phillips)

Disraeli's home at Hughenden Manor, Buckinghamshire, where he
lived from 1847 until his death in 1881 is virtually unspoilt, with the
library and reception rooms lit very much as they would have been.
There has been no need to make changes to enhance the original
lighting, other than to change the source of power from oil lamps to
electricity.

All the above buildings are of special historic value, where the
daylighting has been well handled. While any substantial changes to the
interior lighting could deny the visitor a true experience of its historic
past, cases like this must be treated with special respect. The ruling
principle here is that any artificial lighting which is added should be in
keeping with the period of the property, and that nothing should be
added which could not be removed without in any way damaging the
surfaces of the structure. This is particularly the case with decorative
ceilings.

(a)

(b)

**Figure 3.5**
Saltram House. (a) The Adam Saloon and (b) the morning room by candlelight. The rooms at Saltram are lit in this way to illustrate how they would have appeared in the days of George II. (Copyright The National Trust)

The flame of 'real candles' gives an added dimension, and in the restoration of the original appearance of the George II rooms at Saltram House, Devon, the presence of flame sources has to be experienced. Due to safety regulations the candles are used only on special occasions.

One of the biggest difficulties experienced in putting this book together is to find suitable illustrations of original spaces lit with flame sources. While illustrations of daylit spaces abound, authentic illustrations of the same room at night are rare. Pictures that are available of rooms in our historic buildings are generally taken by means of floodlights or flash, and cannot for this reason represent the true atmosphere of the time. This unfortunately is the case both during the day and at night. All the illustrations in this section demonstrate buildings retained in their original form, and where no 'enhanced' need exists.

## The building retained in its original form and use but adapted for present-day needs

This is by far the largest category, since it includes all the great cathedrals, many churches, temples and shrines, places of assembly, such as the Houses of Parliament in London, university dining halls and libraries and many other building types. This category presents a challenging lighting problem, both now and into the future, as more and more of our historic buildings are refurbished for modern use. The aim will be to allow these fine buildings to continue to serve their original use, but a use which must meet the demands of present-day standards. It must not, however, be confused with those other historic buildings where the use has completely changed and which are described in the final category. There, a different philosophy will be more appropriate.

It will be useful to outline some general principles, and to pose some questions when approaching upgrading problems of this sort:

1 *What needs to be done when considering the enhanced lighting of this category of historic building?*   The first step must be to analyse the lighting requirement from both a functional and an aesthetic point of view. This necessitates the writing of a 'lighting brief' which must be agreed between the client and the lighting designer. No work should be put in hand before this is thoroughly investigated and agreed between all those who have a responsibility for the building. It is of importance also to involve all those who will have to maintain and care for the building with a view to conservation of works of art and building structure.

2 *Does the existing lighting need to be changed?*   What is wrong with it? Can the existing light fittings be modified to fulfil the enhanced needs of the building, without appreciable disturbance to the visual appearance of the space? There are two aspects to this: the light fitting and the light source – the lamp. Modern light sources are much smaller and produce more light, which may make it possible to incorporate them within existing fittings, but, on the other hand, the fittings themselves may need to be remodelled. It should be emphasized that this approach is not the easy option. It requires careful thought and scholarly detailing of the artefacts and cheap imitations will not suffice. Expert advice needs to be sought on the manufacture of such fittings and the way in which they may be modified to achieve the new demands to be made upon them, particularly in the choice of the light source, to provide the required distribution and colour of light. Related problems of overheating need to be considered.

3 *What are the lighting standards that should be applied?*   In many cases this may be obvious, such as the visual tasks in a library. But in some cases, as, for example, in the relighting of a cathedral, it is essential for the lighting brief to be quite specifically related to the individual circumstances of the building so that the regrettable tendency of applying overall solutions is firmly resisted. It may be that the enhanced needs to be provided will necessitate higher levels of light. However, there is a danger of specifying too much light rather than less, where less may be quite adequate for its purpose, and also may add to the magic and enchantment of the building by contrasts of light and shade.

4 *How should the lighting designer approach the problem?*  Every building is unique, and must be approached in terms of an individual lighting problem. There are no standard solutions. The designer should start without preconceived ideas. Also it is important to understand that there is no one 'perfect' solution, there may well be many different yet unique solutions. It will be the job of the designer to start off with an open mind, to consider every aspect of the building, and to build up a solution leading to an architectural whole or 'unity'. Every proposal must be tested in terms of what alterations have to be made to the building structure to accommodate them, and, whether these alterations are irreversible, and if they are, whether they will enhance rather than destroy the spatial fabric.

The most important aspect will be the architect's 'concept' for the overall totality of the space. In this the lighting designer may have a large part to play. Often there may be no architect involved and it will be up to the lighting designer to evolve the initial concept for discussion. A useful approach is to try to put oneself in the position of the original architect or building designer; to ask oneself what he would have done, had he had at his disposal, the modern light sources and technology now available. This is not the same as feeling able to use modern technology as though the original architect had not existed. He should be sitting on your shoulder. He is, after all, the most important part of the equation. The question to ask is 'Would he be in sympathy with the new proposals, bearing in mind the manner in which he originally conceived the atmosphere and spaces within his building?'

It can be argued that the original architect of, say, a church, is only a part of the story, as many subsequent changes may have been made; screens may have been installed separating the choir from the congregation, ceilings may have been dropped over certain areas, side aisles or chapels added, windows enlarged, roofs raised, and so on. It will be the present-day architect's duty to make decisions as to whether such changes should be retained; it is the role of the lighting designer to work within the present architect's solution. The ideal is when the architect is his or her own lighting designer, but as few architects fulfil this role, the case may be made for the lighting designer to have studied architecture. This category of problem, as already mentioned, is the most common and also I believe the most difficult for the lighting designer to solve. Really successful results are not easy to find.

The chapel at Stonyhurst School has daylighting from tall side windows, but the artificial lighting had been found to be gloomy and depressing. Little light reached the ceiling, and the walls were dark. There was little emphasis at the altar and the lighting was unable to fulfil the needs of a modern school chapel, those of concerts, recitals, and theatre, while the old needs had hardly been met.

Changes to the decorative surfaces were made to lighten the interior and the original pendant fittings which had provided a poor overall downward light were replaced with bowls giving upward light to the ceiling. Downward light was provided by concealed directional sources recessed into the ceiling, and special emphasis was given to the altar from concealed sources at high level. Although the changes were carried out twenty-five years ago, the results are still satisfactory.

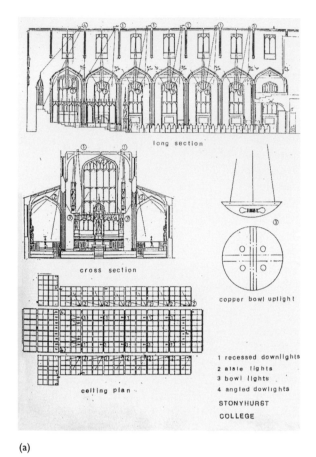

(b)

(a)

**Figure 3.6**
Stonyhurst School. (a) Design of the chapel lighting in plan and section with details of the upward lighting fitting. (b) The chapel, showing the overall effect of the completed lighting scheme looking towards the altar. (Copyright Derek Phillips) (Architect for the restoration, Building Design Partnership; lighting consultant, DPA Lighting Consultants)

At Warwick Castle some of the rooms have been made into historic 'sets' designed to give the public an impression of the appearance of the room as it would have looked when used in the past, depicting well-known figures such as Sir Winston Churchill, thus avoiding the rather 'dead appearance' of many unoccupied historic buildings. The display lighting used is well hidden from public view. This is easy to do, when as in this case the view is unilateral.

(a)

(b)

**Figure 3.7**
Warwick Castle. (a) Lady Warwick's boudoir and (b) the music room with waxwork figures depicting a royal weekend party at the castle in 1898. (Copyright Warwick Castle)

## The historic building for which an entirely new use is planned

Many historic buildings are being given a new lease of life by a 'change of use'. Examples range from a stately home being used as an hotel, an old factory as an art gallery, and a market as a modern computer centre.

The essential aspect of this type of problem is that once the decision has been made to change the use of the building, there must be a general presumption that significant changes are likely to be required to the overall planning of the building. The lighting design can then employ the most up-to-date technology, and when this becomes outdated it would be appropriate to change it yet again.

This is a different approach from that outlined for the building with an enhanced use, where the aim is to alter as little as possible, and in the view of English Heritage 'nothing being done which cannot be undone.' Such an approach would set the lighting for the building in tablets of stone. Having said this, the normal rules applying to the relationship of light and architecture will still apply, and many of the questions already posed will still be relevant.

As in new buildings, the spatial functions need to be investigated and fully understood. The difference is that in an historic building the existing structure will impose a degree of discipline on the solution, particularly in terms of its daylighting, so that, effectively, the historic building will have imposed its own character on the architecture and consequently on the lighting solution. A visitor entering the building should be well aware of the fitness for purpose of the new interior, but for the design to achieve real success the visitor should also be conscious of the historic nature of the structure. A good example of this is illustrated by the designs for the Bankside Power Station, London, to be developed as additional space for the Tate Gallery (see Chapter 7).

It is not the purpose of a book on lighting to try to deal with the difficulties this represents for the architect, but rather it should be said that this will offer a great opportunity for the architect to demonstrate ingenuity and skill. This must be supported by a lighting design of similar quality, in which the most up-to-date technology is coupled with

**Figure 3.8**
Musée d'Orsay, Paris. The Gare d'Orsay was converted into an art gallery in 1984–1986. The daylighting of the original trainshed still provides the main environmental light during the day. (Copyright *International Lighting Review*)

a knowledge and understanding of the integrity of the original structure. This is no easy task.

An attempt has been made in this chapter to analyse the different problems presented by the lighting of historic buildings; to show the need for an understanding of their nature and function, in terms both of their original context and of today. Whether the fundamental use of the building remains or whether it is changed, the designer has a challenge to face.

## INTRODUCTION TO FOLLOWING CHAPTERS

The purpose of the following chapters is to discuss and illustrate different types of historic building; to show how the building would have been lit in the past, and then how it has been adapted to present-day use. The three types of lighting problems analysed in this chapter will be reflected in the nature of the solutions developed, differing from the building enjoyed purely for its historic value at one extreme to the building for which an entirely new use has been agreed at the other.

It is not suggested that there is only one 'right' way to solve the lighting of historic buildings. There are indeed many ways and there is no wish to put the architect or lighting designer in a straitjacket. Perhaps it would be more appropriate to say that evidence shows that there is an infinite number of 'wrong' ways, which have been selected in the past either from ignorance or economic expediency, and my purpose is to direct the designer away from such solutions.

To do this a logical design approach must be adopted, an approach which must adapt to the different circumstances of the individual programme. To assist the reader a series of conclusions is added at the end of each chapter.

# 4  Domestic buildings

While flares or oil lamps were used by the original cave painters 4000 years before Christ to decorate their environment, the history of the lighting in peoples' homes is synonymous with the history of day-lighting, with flame sources used initially as functional elements to provide security and warmth. It is doubtful if the 'lit' appearance of a space after dark was considered to be of importance much before the Greeks and Romans created villas, where the floors and walls were decorated with paintings or mosaics, and where the suspended oil lamps, such as those in Pompeii, permitted the interior to be seen and enjoyed for its own sake. From here on, 'domestic lighting' would have had social significance, the degree of lighting in homes expressing their owners' affluence.

The approach to lighting in historic houses will depend upon the use to which the building is to be put. The simplest approach is where the house is still used for living in, and where the lighting must meet the everyday needs of the occupants. The design must reflect the age and character of the dwelling, whether a medieval manor house or a Palladian mansion, with the infinite variety of residential types in between. The lighting of the manor house would have been very primitive, consisting of candles, although the finest were expensive, or oil lamps, while the Palladian mansion would have had suspended chandeliers with multi-bracket candles and wall brackets. Both no doubt would have had various forms of portable lamp, which could be used in the main living areas or, alternatively, carried around the house to illuminate parts which would otherwise be unlit. The emphasis must be on satisfying the needs of the occupants, while respecting the integrity of the architecture.

A different approach must be applied to the residence which is now exploited for tourism, and where visitors come as much to see the artefacts, such as paintings, carpets, or furniture, as to enjoy the beauty of the house itself, an enhanced use. Each problem needs to be analysed to establish the importance which should be given to the house and to the objects within. The danger is that in order to see the artefacts, special lighting arrangements may need to be made which could lessen the impact of the house.

A delicate balance needs to be struck, but I believe that where the architecture is of quality (which will no doubt be the normal case) the appearance of the house should not be prejudiced by its contents. The ideal must be to make it less of a museum, more a 'lived-in' home.

## DESIGN APPROACH

In both the case of the house for 'living' and the house as 'display', the design approach must first be to study the daylighting, since this is where the design for the interior will have begun. The orientation of the spaces will be of importance – which rooms benefit from sunlight, and at what times of day, the modelling of the interior by shadow patterns and where for reasons of conservation it may be a requirement to reduce the quantity of daylight, some or all of the time. There is no imperative to echo the daytime appearance after dark. The house should take on its own intrinsic quality at night. It will be a different experience, as it should be, but one that should be equally valid.

Next, it is necessary to establish an 'artificial lighting' brief. There will generally be some form of lighting – candelabra, wall brackets, etc. – and account should be taken of these to see to what extent they are correctly positioned and whether modern technology can improve them. The tendency will be to follow the original locations and types of light, although the latter may not necessarily be correctly sited. The principles of good home lighting should be employed, to ensure that light is provided where necessary for the different functions in the house – living areas, dining rooms, kitchens, circulation spaces and bedroom/bathroom areas.

Where the brief is for the lived-in house, this, together with the designer's sense of history, should enable a lighting scheme to be prepared which will meet the needs of the occupants without destroying the integrity of the structure. Portable fittings can be very useful in living areas, to provide the necessary ambience together with the higher levels of light required for household tasks such as sewing or reading, without glare to others. Glare kills atmosphere, while making it difficult to carry out normal tasks. It is always better to have a number of smaller light sources which are well shielded rather than fewer and brighter lamps which may cause glare.

Where the property is open to the public the brief has to become more sophisticated, and decisions must be made as to what the visitor needs to see in addition to the house and overall environment of the property. An inventory should be made for each room to include the artefacts, furniture, special architectural features, wall surfaces, paintings and so on, together with the desirable quality and colour of light required for each item. Indeed, this will be helpful in identifying groups of objects with similar characteristics.

A brief can then be provided to indicate the nature of the light required. For example, where important paintings are placed in a room, these may require individual attention, while the less significant works may derive their light from the overall building lighting. The advice given by a director of an art gallery when asked how to light a painting was 'put it in a well-lit room'. He might have added 'provided that the needs of conservation are met'.

There is a problem in obtaining credible illustrations of the lit appearance in early buildings of flame sources as the original lighting will generally have been removed, and it is difficult to reproduce the original effect. When William O'Dea was embarking on his excellent book *The Social History of Lighting* he felt it desirable to find old lamps for a variety of oils, have special candles made, and to experiment with the characteristics of early light sources. However, his experiments appear to have been limited to the 'management' of these early light sources, and to their economics, rather than to their contribution to the visual appearance of a space, and there are no illustrations of the rooms in which they were placed. We are therefore in the most part limited to the appearance of the light sources themselves.

It is difficult also to rely on the commercial photographs of interiors which may be purchased from the houses themselves. While they are beautiful pictures, they are for the most part taken with fill-in floods, and tend to embellish rather than give a true picture of the lit interiors. Although it is difficult to find illustrations of the original lighting in buildings, there have been some excellent reconstructions in motion pictures, for example, the film director Stanley Kubrick had special candles made to allow him to shoot scenes in *Barry Lyndon* without the use of extra flood lighting, giving an authentic eighteenth-century experience.

There now follows a description of a number of houses which represent some of the different periods of architecture and the approach that has been made to the lighting, both daylight and artificial.

## BAYLEAF HOUSE, SUSSEX

A Tudor farmhouse built originally in 1536 and re-erected at the Weald and Downland Museum in Sussex gives a good example of how the spaces would have been lit. By day, open 'window' openings provided daylight, but as there was no glass, they were closed by shutters at night, or on the side facing the prevailing wind during the day when necessary. A fireplace in the centre of the hall provided heat, warming facilities for food cooked in the kitchen and light, the smoke being allowed to filter out through the roof. In addition to the central fireplace, candles were used after dark around the walls, while candles alone were used in the bedroom areas above.

(a)

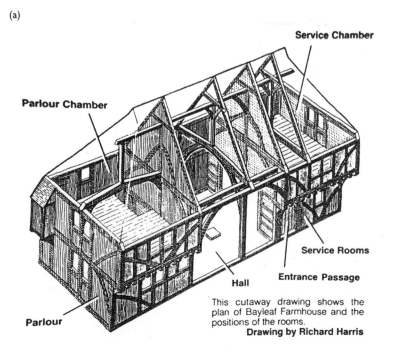

Service Chamber

Parlour Chamber

Service Rooms

Entrance Passage

Hall

Parlour

This cutaway drawing shows the plan of Bayleaf Farmhouse and the positions of the rooms.
**Drawing by Richard Harris**

(b)

**Figure 4.1**
Bayleaf House. (a) Cut-away construction drawing, by Richard Harris. (Copyright Weald and Downland Museum.) (b) Exterior of the house. (Copyright Derek Phillips)

(c)

**Figure 4.1** (cont.)
(c) The main hall during the day, with windows open to the elements. (d) The hall lit at night by the central fire and candles. (e) The parlour chamber/bedroom during the day. (f) The parlour chamber/bedroom with shutters closed at night and lit by candles. (Copyright Derek Phillips)

(d)

(e)

(f)

## KENTWELL HALL, LONG MELFORD, SUFFOLK

This moated Tudor hall, built in 1560, has been restored by its present owners over the past 20 years, and acts both as their home and as an attraction for special parties of visitors who wish to enjoy the impression of living in Tudor England. The house crosses the boundary between being a private one for family use and a tourist centre, as a real attempt has been made to give it 'life', unlike many 'stately homes' which are unoccupied and have a 'deathly' appearance.

The house has been refurbished with as much authentic furniture and artefacts as possible, and the artificial lighting is designed to be used principally for use by the family. The visitors enjoy a more authentic experience than if forms of display lighting had been employed. The philosophy was that if the lighting is suitable for the family at home, then, with minor exceptions, it should be suitable for the visitors.

(a)

**Figure 4.2**
Kentwell Hall. (a) The exterior of the Hall by day. (b) The Great Hall by daylight. Bilateral daylight gives good-quality light. (c) The Great Hall by candlelight at night. Some wall-mounted candles give light, otherwise lighting is at a low level from portable candles.

(b)

(c)

**Figure 4.2** (cont.)
(d) The kitchen by day. (e) The kitchen by night, with modern lighting added to light the ceiling at high level for the benefit of visitors. (Copyright Derek Phillips)

## TEMPLE NEWSUM, NEAR LEEDS

Owned by Leeds City Council and run by them as a part of the City Art Galleries, Temple Newsum has become a powerhouse for the historic study of aspects of great English houses. The Council published the fourth in a series of Country House Studies, entitled *Country House Lighting 1660–1890* gathering information and light fittings from many different houses. The house was used as an exhibition itself, and the exhibition then travelled to other locations.

What is most interesting is that the house today is an authentic example of a residence, albeit a very large one, in which the artificial lighting takes its rightful place with other aspects of the decoration. It is not a museum, the house itself is the exhibit. All the light fittings are genuinely of the period, early sixteenth century although they may have been gathered from various locations. They represent what a contemporary would have experienced.

(a)

(b)

**Figure 4.3**
Temple Newsum. (a) Exterior showing the window layout. (b) Detail of typical windows from outside.

**Figure 4.3** (cont.)
(c) The Picture Gallery by daylight. This was described by a contemporary as 'an immense gallery hung with red and covered with pictures. . .the light from the fire and the lamps gleamed on a little tea table and a few chairs around it, all beyond was lost in dark immensity. . .it was like arriving at a bivouac in a desert'. (d) Girandole wall bracket for candles. (e) The first-floor 'oak' corridor, creating a 'place'. (f) The Georgian Library by daylight.

(c)

(d)

(e)

(f)

**Figure 4.3** (cont.)
(g) The Great Hall by day. (h) The Great Hall by night. (i) The dining room by day. (j) The dining room by night. (Lighting consultants the Temple Newsum curatorial staff; copyright Derek Phillips)

(g)

(h)

(i)

(j)

# CONVERTED STABLES, HERTFORDSHIRE

An example of simple lighting applied to the bedroom suite of a converted stables dating from the middle of the nineteenth century is an old barn in Hertfordshire. This is illustrated by a series of views indicating the atmosphere created by light at different times of day and with different light sources.

The solution to the lighting of this room has adopted the following principles. The daylighting, although from low windows, is adequate, assisted by a light-coloured floor. The artificial lighting is simple, effective and cheap to buy and run. It suits the bare functionalism of the original space. Both daylight and artificial light are adequate for their purpose.

**Figure 4.4**
Converted stables, Hertfordshire. (a) Plan and section of the bedroom in converted first-floor barn. (b) The bedroom in morning sunlight, received from two side windows, overlooking an external courtyard.

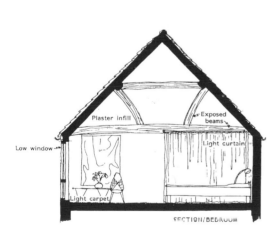

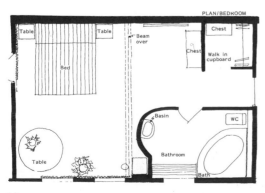

(a)

(b)

(c)

(d)

(e)

**Figure 4.4** (cont.)
(c) The bedroom in bright daylight, showing the ceiling well lit by reflections from the light-coloured carpet. (d) The bedroom at night lit by two 300 watt tungsten halogen lamps placed on top of the low cross-beam. (e) The bedroom lit by both uplight and bedside table lamps.

(f)

(g)

(h)

**Figure 4.4** (cont.)
(f) The bedroom lit only by bedside lights, the ceiling is relatively dark. This is a suitable light for reading in bed. (g) The bedroom lit by candlelight to achieve the same effect as when it was used in the nineteenth century. (h) A view of the two windows from which the daylight to the room is received. (Architect and lighting consultant Derek Phillips; copyright Derek Phillips)

## BENENDEN SCHOOL, KENT

Benenden is one of the finest girls' independent schools. Built in 1862, it was originally the home of the Earl of Cranbrook but was sold to Viscount Rothermere in 1912, who undertook a major reconstruction. The building has been used as a school since 1924. In order to keep up with modern developments in education the school has completed a scheme of reconstruction, with additional modern accommodation.

This new construction has prompted the reconsideration of the existing buildings to ensure that they too provide an excellence of environment. This presented a good opportunity to upgrade the internal lighting in the main building. Little of the original lighting for the historic house remains, and functional fluorescent light fittings had been installed, both as pendants and as uplighting around the friezes. While the level of light for functional needs was not altogether unsatisfactory, the visual appearance of the fittings was often glaring, and did little justice to the quality of the spaces. The aim was to recreate the original residential atmosphere of a boarding school.

Consultants were appointed in 1993 to reappraise the lighting, bearing in mind the need to enhance the decorations, and at the same time provide functional and energy-efficient solutions. The main entrance and foyer, used by parents and visitors, were lit almost entirely by two small chandeliers, creating a gloomy atmosphere. The space has double height at the staircase, giving access to the headmistress's room and dormitories above, with a decorative strapwork ceiling, while there is a lowered ceiling at the entrance end. The effect sought was to combine the furniture and layout with the lighting design to recreate the atmosphere of welcome one might find in the lobby of a first-class hotel.

One chandelier was removed, giving emphasis to the remaining one over the sitting area. The plain lower ceiling incorporated miniature low-voltage recessed downlights to highlight the entrance area, while the upper decorative strapwork ceiling over the staircase was uplit to enhance its colour and pattern using three surface-mounted metal halide uplights. Table and floor standards were then related to the sitting area and the monumental fireplace.

The first-floor study was originally Lady Rothermere's boudoir, and now serves as a drawing room used for small meetings. The lighting design provides light to the three-dimensional decorative frieze, even during daylight, when the room is sunlit. After dark the emphasis changes to focus on the fireplace and meeting table. The frieze and linenfold panelling ring the room with light, instead of being lost in gloom.

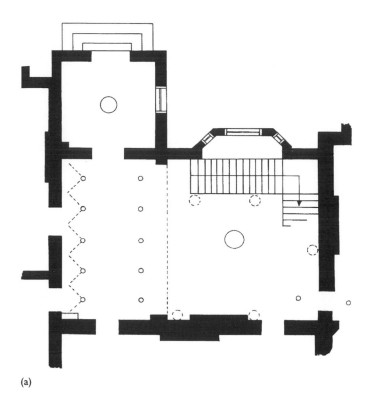

(a)

**Figure 4.8**
Benenden School. (a) Plan of lobby/foyer to show the lighting layout. (b) The lobby/foyer seen during the day. The original chandelier has been retained over the staircase area associated with the table and floor standards. (c) The lobby/foyer at night looking towards the lower entrance floor area lit by miniature low-voltage downlights recessed in the ceiling. (d) The drawing room, formerly the headmistress's study, now used for small meetings. (Architect David Brander; lighting consultant DPA; copyright Derek Phillips)

(c)

(b)

(d)

## HERTFORD HOUSE, LONDON

The original house, which was altered and extended between 1872 and 1875, by the architect Thomas Ambler, was the London home of Sir Richard Wallace, later Lord Hertford, and housed his collection of paintings, furniture and other artefacts. Therefore, rather like Sir John Soane's Museum, it was both a residence and a collection which is now open to the public. The house is built around a central courtyard, with stables and servants' accommodation forming the outer three sides at low level, with galleries down the long sides above.

The Reynolds room, used by Sir Richard as a study, was lit from a large overhead candlelit chandelier, with other candlelit wall brackets and candlesticks on tables and chimneypiece. Gaslight had been available from the early 1800s, and was installed at the house in the late nineteenth century, being converted to electricity in the early twentieth century. The lighting in the long gallery was originally by means of overhead daylighting through a laylight, but this has now been closed in above with the light derived from electric sources.

**Figure 4.9**
Hertford House. (a) The courtyard today, with the stables no longer in use and the entrances filled in. Similarly, the galleries above are closed off. (Copyright the Trustees of the Wallace Collection)

(b)

(c)

(d)

**Figure 4.9** (cont.)
(b) The Reynolds Room with wall brackets and central chandelier. (Copyright the Trustees of the Wallace Collection.) (c) The Long Gallery with the large laylight originally admitting daylight through an overhead laylight. (Copyright the Trustees of the Wallace Collection.) (d) The Long Gallery today with the laylight now lit by electric sources and the centre filled with illuminated domes. (Copyright Derek Phillips.) (Architect Thomas Ambler; lighting consultants the Wallace Collection staff)

## CRAGSIDE, ROTHBURY

The Victorian home of Sir William Armstrong, an engineer and industrialist, later Lord Armstrong, was started in 1863 as a modest two-storey lodge for shooting parties, but in 1869 he appointed the architect Norman Shaw to enlarge it. At the same time Armstrong was purchasing large tracts of land to enable him to control the water supply by means of damming a lake. He had a fascination with hydraulics, and later used hydraulic power from his own estate to power a generator to create electricity for help with estate management and, more importantly here, for lighting.

Craigside was the first house in the world to be lit using the new electric light bulb invented by Armstrong's friend Joseph Swan. He used 45 lamps, 20 on the gallery alone. Most of the grand houses of the day were now lit by gaslight, and for some years there was much competition from the two sources of power for light. Armstrong's use of electric light throughout Cragside set the standard by which later houses were to be judged, and, apart from the efficiency of the light sources now available, little has changed. But, as the design of the windows show, daylight was still considered to be the most important aspect of residential lighting.

**Figure 4.10**
Cragside. The Library. A well-daylit room, but the new 'electric' pendants installed by Lord Armstrong can clearly be seen. (Architect, Norman Shaw; photographer, A Von Einsiedel; copyright the National Trust)

(a)

(b)

(c)

(d)

**Figure 4.11**
Waddesdon Manor. (a) The Blue Dining Room in daylight showing the beautiful crystal glass chandelier unlit. (b) The Blue Dining Room at night. This shows the lit chandelier together with the candles which would normally have been lit for dinner. (c) The Morning Room in daylight which shows the interior with its many artifacts and furniture that give a 'lived-in' impression. (d) The Morning Room after dark with its chandeliers and picture lighting, the latter being very subtle showing the paintings to advantage without glare or unpleasant reflections.

# 5 Ecclesiastical buildings

The lighting in early churches consisted mainly of the daylight available in different parts of the world, the structure of the building reflecting the external climate. There were small openings where light was plentiful, while whole walls of light were planned for northern latitudes where daylight was scarce.

Christian churches were designed to allow early morning daylight to be received from the east with south light flooding into the naves and the southern aisles during the day. In the evening, light from the west would enter from behind the congregation. All the great cathedrals in the northern hemisphere worked on this principle. The fact that early morning light entered from the east or from behind the altar meant that the altar would have been seen in silhouette against the brightness of the east window, so that the contrast with its background would have made it difficult to see once the east light was at all bright, and the sun had risen. For this reason, the early offices of the Church would have taken place before sunrise. This principle applied not only to the great churches but universally down to the small parish church, so that the orientation of the traditional church was on an east/west axis, around which the ritual of the church developed.

After dark the situation was quite different, with what flame sources were available confined to the immediate vicinity of the altar or other aspects of the ritual. As with the little cathedral in Iona flooded by natural light during the day, but achieving a totally different appearance by candlelight at night (Figure 3.1), so the night lighting of Cartmel Priory, Cumbria, indicates a dramatic difference between the appearance of daylight and that derived from the artificial sources at night.

The manner in which the flame sources were supported was basically functional, in terms of its location, whether suspended at a low height where candles could be reached and lit, alternatively mounted low on walls, or related to items of furniture. The functional aspects were not permitted to compromise the desire for decoration, and church fittings were often highly decorative, as depicted by the Gloucester candlestick of the twelfth century illustrated in Figure 2.2 (d).

There are few churches of whatever denomination where the function has not changed since they were first built; for reasons of changes in the church ritual itself or because the church is now used more extensively or at different times of day. Often there are new uses for the church, such as education, concerts and musical recitals, or a theatre. The original lighting would no longer be adequate or have sufficient flexibility to cope with the

(a)

(b)

**Figure 5.1**
Cartmel Priory. (a) The interior lit by a combination of high-pressure sodium and tungsten halogen lights. (b) An alternative view of the interior. (Lighting consultant Robin Wright in association with James Bell; copyright James Bell)

enhanced role while still performing its primary function as a place of worship. Change has often been led by the need to move from one light source to another (gas to electricity) or because the original electric wiring needed replacing.

For whatever reason, the lighting system may need to be changed, and this needs careful thought. Chapter 3 sets out a general theory for the lighting of historic buildings. This is the starting point, but there are some aspects of church lighting which need special emphasis.

First, the structure of the building must not be compromised either structurally or visually. Old churches are generally formed of solid masonry, where no allowance would have been made for concealed pipes or wires. Where it is necessary to apply new light fittings to solid masonry, the route to be taken by the electrical supply must be a first consideration to ensure that nothing can be seen to disturb the appearance of the interior architecture.

> The electrician's artistry,
> His Clapham-Junction-like creation
> Of pipes and wires and insulation
> Of meters, boxes, tubes, and all
> Upon our ancient painted wall.

(From 'Poems in the Porch' by Betjeman, 1958)

Many old churches are spoilt by the addition of inappropriate light fittings which mar the daylit appearance of the building, and, where possible, light sources themselves should be concealed from normal points of view.

Having said this, there has been a tradition in the past for lighting equipment manufacturers to publish catalogues of 'church ranges' of pendants. It is easy to see the logic in that, once installed at low level, the pendants may easily be maintained. Unfortunately the design of such fittings often left a lot to be desired.

There may be reasons for the introduction of well-designed pendants in historic churches, light sources being related to materials of metal or glass, adding a positive element to the interior design. Such chandeliers should be placed at such a height as not to impede the views of the congregation, and where the lamps are concealed to obviate glare from normal points of view. Sir John Betjeman quotes Sir Ninian Comper's advice as suggesting 'as many bulbs as possible of as low power as possible, so they do not dazzle the eye, when they hang from the roof'. This is taken to mean that sparkle rather than glare is permissible.

Second, an understanding of the building, its plan and section and the way this has contributed to its function, and how it is intended to be used in the future, will relate to its date and the nature of its architecture, whether Saxon, Norman, Gothic, or the later styles of Renaissance and Baroque. All were developed within the discipline of daylight, the window types reflecting the available technology. The most obvious examples of this are the tall gothic windows of the twelfth and thirteenth century cathedrals, where the side walls became 'window walls' to admit the maximum daylight.

One does not have to be an architectural historian to understand the nature of the different styles of church, but it does require a certain sensitivity. The essential static verticality of the Gothic, contrasts with the more horizontal movements of the Renaissance and Baroque.

Third, there are the unique qualities of night-time lighting. There is no reason why the appearance of a church at night should be the same as that during the day. While it must appear to be the same space, there seems to be no good reason which determines that it should look the same (even if this were physically possible) and there are many reasons why it should not. Our expectations of a church at night are different. There is the variety of experience to be added after dark to the already rich variety of mood created by the changing exterior. At night it may be appropriate to provide a degree of theatre lacking during the day.

At St Albans Abbey in Hertfordshire on Christmas Eve, the entire congregation is given a small candle on arrival for the service, and at a certain point, the artificial lighting is lowered and everyone invited to light their candle, after which all the artificial lighting is turned out. The effect of this massive array of candles is magical.

Jack Waldram, one of the most famous lighting engineers, when lighting Gloucester Cathedral in the 1960s, believed that the light at night should reflect daylight distribution, so he arranged a preponderance of artificial light from the south side at night to light the north walls to a higher degree than the south, in order to echo the directional flow of daylight. This seems contrary to the concept that the night environment should have its own intrinsic quality and today may appear to be 'artificially' artificial.

Fourth, while it is desired not to waste energy, there is less pressure to use the most energy-efficient equipment in a church, where the lighting may be used for only a small part of the day. What is more important is that the light sources to be used are capable of dimming control to satisfy the needs of flexibility, and at the same time save energy. However small the church, the capital expenditure on a 'scene-set' control system, allowing the lighting to be changed from one part of the service to another, should be considered.

As all church lighting problems are different, it will be useful to discuss a number of examples from different periods of architecture to see the results of the work of the lighting designers, and to understand how they arrived at their solutions.

(a)

(b)

**Figure 5.2**
Gloucester Cathedral. (a) Sketch by the lighting designer Jack Waldram to illustrate his concept for the relighting in 1957. (b) The interior of the nave at night, showing the flow of light from the south side towards the north. (Copyright Derek Phillips.) (Lighting consultant J. M. Waldram)

## DURHAM CATHEDRAL

Durham has one of the finest medieval cathedrals in Europe. It was built after the Norman conquest of Britain, between the years 1096 and 1133, a comparatively short period of 40 years for such a massive structure.

Artificial lighting had been installed in 1930 and the cathedral was rewired in 1970. A decision was made in 1990 to update this installation, and consultants were appointed to prepare proposals. The previous lighting had concentrated the light from high level onto the pews. This had proved to be glaring and did not reveal the beauty of the architecture. So the brief for the redesign was as follows:

1 To improve the general lighting for the congregation, while eliminating glare
2 To improve the lighting to those areas of the cathedral used as the focal points for the services, the sanctuary, the crossing, the feretory and the west end
3 To highlight certain architectural features, such as the ceiling vaulting and the Neville Screen, of interest to visitors, and to ensure that the proposals paid due regard to economy in use.

Trials were carried out on-site to demonstrate to the cathedral authorities the relative merits of two alternative light sources: metal halide or tungsten halogen. The latter was selected despite its lower efficiency, due to its superior colour rendering of the 900-year-old stonework, and not least because of its ease of control by dimmers, allowing greater flexibility. The energy issue is less crucial in a church where use is more limited, and sources can be either off or at low power for much of the time.

By choosing one type of luminaire which can be adapted to provide different distributions by the addition of alternative lenses the installation was simplified. Most of the equipment is mounted at upper clerestorey level, where it can easily be concealed to light downwards to the congregation or across to light the walls or triforium and upwards to the vaulted ceiling, as far as possible minimizing glare from the lamps.

The whole of the lighting system is controlled by a 24-scene programmable dimming system, enabling lighting emphasis to be placed where required to meet the changing needs of the cathedral, on the one hand, congregations of a thousand or more, and, on the other, quite small groups. The needs of a cathedral have greatly expanded to encompass activities such as concerts and theatre and the lighting must be sufficiently flexible. It was the designer's intention to 'balance the various elements' to avoid the domination of one element (such as the ceiling, as has often been the case) over the rest, while allowing special emphasis on a particular part, such as the Neville Screen. The Neville Screen has cross-lighting from closely controlled spots at high level, which emphasize the modelling of its ornate Caen stone sculpture.

**Figure 5.3**
Durham Cathedral. (a) The nave by daylight. (b) Artificial light added to the nave during daylight. (c) The nave at night. (d) Night view showing the Scott screen. (Lighting consultant and copyright Lighting Design Partnership)

(a)

(b)

(c)

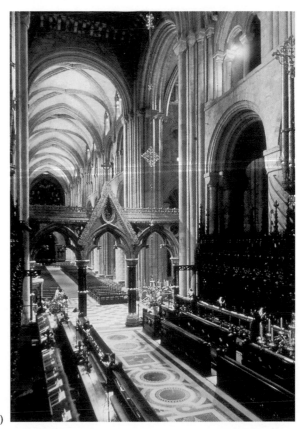

(d)

## UDINE CATHEDRAL, ITALY

Built in the thirteenth century, Udine Cathedral has three parallel naves
and is richly decorated with frescoes. The lamps chosen to light up to the
ornate ceilings of the three naves and downwards to the floor are metal
halide. These are placed at the top of each column at the junction between
the springing point of the arches above, 400, 250 and 100 watt lamps
being used to obtain the desired distribution. The downlights have a
strictly controlled distribution to eliminate glare to the congregation. (It is
interesting that at Durham the tungsten halogen lamp was selected
instead of the metal halide, due to its dimming capability.) The side aisles
and chapel use the warmer high-pressure sodium lamp (SON), with a
similar system of upward and downward light. The change in light
source gives a contrast between the cool, high interior and the warm,
lower chapels.

The metal halide lamp was selected due to its high efficiency. The
whole system is controlled by a microprocessor unit capable of 36
different combinations, allowing the nave to achieve a desirable balance
of upward and downward light, reacting to differences of climate and
season, with a change from daylight to night-time.

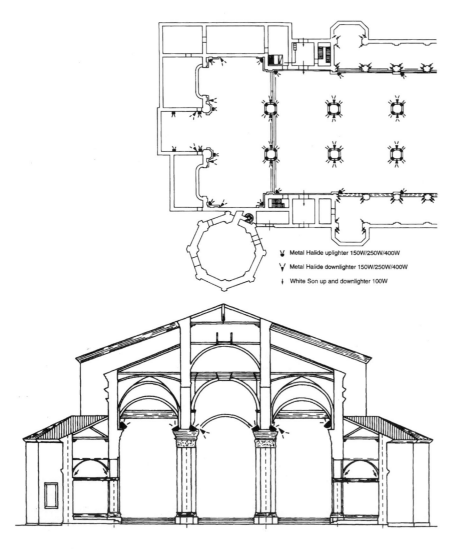

**Figure 5.4**
Udine Cathedral. (a) Plan and section to show
the lighting scheme.

V̌ Metal Halide uplighter 150W/250W/400W

V̂ Metal Halide downlighter 150W/250W/400W

✦ White Son up and downlighter 100W

**Figure 5.4** (cont.)
(b) View of the nave to illustrate the lighting of the vaulted ceiling from the tops of the columns.

**Figure 5.4** (cont.)
(c) View towards the side chapels, which illustrates the contrast between the warmer White SON light with the cooler metal halide in the nave. (Architect S. Daffara; lighting consultant Philips Lighting; copyright *International Lighting Review*)

## HOLY TRINITY PARISH CHURCH, COOKHAM, BERKSHIRE

The parish church at Cookham dates from the thirteenth century and the opportunity arose in 1992 to upgrade the lighting originally installed some fifty years ago. The reasons given were fairly typical of this age of church:

1 The original electric wiring had decayed and required replacing to modern standards.
2 The original electric light fittings located at beam level gave downlight only, leaving the roof in darkness. The fittings had a wide distribution, causing unacceptable glare when facing the rear of the church.
3 New uses for the church include theatrical performances and concerts, demanding a more controllable lighting system where different areas can be emphasized at the expense of others.
4 The church is now a place of pilgrimage, so that there was a need for emphasis on some memorials during the day when in general, daylight would be sufficient.

A lighting scheme was developed which had the following parameters. The nave and chapels should be provided with glare-free and adequate light to enable the congregation to participate in the services, with upward light to the fine timber roof above. Each facility should be separately controlled to permit the desired emphasis to suit the different needs of the church. The fittings were mounted on the cross-beams as before, a practical solution, the downward lights being of a closely controlled distribution to avoid glare. The chancel and altar had a separate circuit of light sources, concealed from view of the congregation and additional sources were provided for special emphasis on memorials.

The daylight appearance of the church does not suffer from the new lighting fittings, which, being black, fit the dark colour of the beam structure. The electric wiring has been well concealed, and this modest and unobtrusive scheme enhances the night-time effect, while 'scene-set' control permits the flexibility requested by the church. A particular requirement was that due to the pilgrimage nature of the church, and its connections with the artist Sir Stanley Spencer, the church should be open for visitors at most times. A special lighting circuit is left on during the day to provide emphasis on the memorials and a painting by the artist. The low energy required to do this is acceptable to the church.

**Figure 5.5**
Holy Trinity Church, Cookham. (a) Plan and section illustrating the general lighting arrangement.

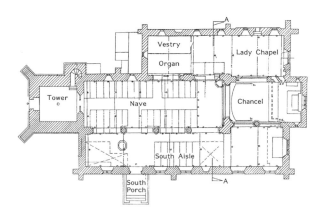

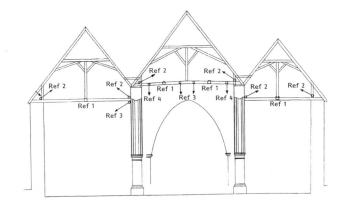

(a)

(b)   (c)

(d)

**Figure 5.5** (cont.)
(b) The exterior of the church. (c) Daylight view towards the chancel, with supplementary artificial light. (d) Night view towards the chancel. (e) The side chapel lit as a pilgrimage church with lowered lighting. (f) The chancel. (Lighting consultant DPA; copyright Derek Phillips)

(e)   (f)

## CHAPEL ROYAL, HAMPTON COURT

Figures 5.6 (a) and 5.6 (b) showing the changes made to the lighting of the Chapel Royal speak for themselves. The main changes are to the choir stalls, reredos and altar.

The original lighting used concealed 10-foot fluorescent lamps behind each column flanking the altar, and the choir stalls had domestic silk shades which obscured the choir master. The music was poorly lit, and the whole subject to glare. The fluorescent lamps and silk shades were removed and the raised choir stalls relit using a small tubular brass fitting supported by a slender tube carrying the electric supply. The choir can see the music and the choir master, and there is no glare from these fittings, allowing a better view to the reredos and altar, which have been relit using concealed low-voltage lamps.

The 'Gloria', the painting above the altar which previously had been invisible, is now lit using a long-throw low-voltage spot, and can be seen well. Finally, the public area below the royal pew, also originally lit by fluorescent lighting, has more appropriate brass chandeliers.

The ornate gold ceiling is now well lit by interreflection, and in addition, since the needs of the chapel are simple, one scene for visitors and a second for use as a chapel, the original complicated lighting controls were removed and a simple switch-over arranged as between lighting for services and lighting for visitors enjoying the chapel.

(a)

**Figure 5.6**
Chapel Royal, Hampton Court. (a) View of the chapel before the relighting was carried out. The 'Gloria' behind the altar cannot be seen. (Copyright Paul Ruffles)

(b)

**Figure 5.6** (cont.)
(b) The completed lighting scheme in 1985. The altar area is much improved as is the lighting to the choir stalls. (c) The choir stall lighting. The new fitting was designed to improve the lighting on the music, as well as the view of the choir master seen by the choristers. (d) The 'Gloria' lit by a long-throw 7° beam from a 50 watt low-voltage dichroic lamp. (Lighting consultant BAP (William Allen); copyright J. Cockayne)

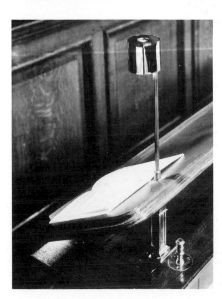

(c)

(d)

## ST MARTIN-IN-THE-FIELDS, LONDON

The original church, built in medieval times, was rebuilt as the royal parish church in the eighteenth century by James Gibbs. It is simple in plan, having a rectangular nave with a tall front Corinthian portico and western steeple. A sanctuary containing the altar is formed to the east, with galleries down each side. It was the prototype for many American colonial churches.

For various reasons, not least because of some structural failure in 1960, the whole church needed to be refurbished, and this, together with the redesigned lighting, was completed in 1980.

Much of the architectural quality of the church lies in the modelling and decoration of the plasterwork ceilings, a large elliptical barrel vault over the central area of the nave, with shallow domes over the galleries. These were not well lit by the natural light which enters during the day through side windows, and the original lighting, consisting of pendant indirect fittings, gave little light below to the pews and a bland shadowless light to the ceiling.

Four large traditional Dutch chandeliers, similar to those originally installed in the eighteenth century were designed to replace these, giving modest upward light and sufficient downward light for the congregation. The chandeliers were designed to provide a degree of sparkle, but without glare. They are placed below the gallery level enabling those in the gallery to look over the top. In addition to enhance the beauty of the main ceiling, it is uplit by two 500 watt tungsten halogen floods from one side to emphasize modelling. These are placed at the springing point of the arches to the gallery, coinciding with the entablature to the columns.

To reassure the church authorities that the scheme suggested would meet their criteria, a mock-up was carried out in one bay. This enabled the architect to judge the quality and colour of the upward and downward light, and how it would reflect off the curved surfaces of the interior. It is difficult to calculate the effect of such lighting proposals and the cost of such a mock-up is a small price to pay if it conveys the lit impression that will be achieved.

Two smaller chandeliers were designed for the sanctuary, matching the larger ones in the nave, and additional concealed spotlights illuminate the altar and the ornate plaster ceiling above. Finally, the saucer domes above the galleries are uplit using spotlights with fittings recessed in the arches above providing direct light to the galleries. The whole system is controlled by a dimmer system, giving a variety of lighting patterns. The lighting below the galleries is by reflected light from the scallop domes using wall-mounted uplighting from 300 watt tungsten halogen floods.

**Figure 5.7**
St Martin-in-the-Fields. (a) The exterior of the church.

(a)

(b)

**Figure 5.7** (cont.)
(b) General view of the nave. This shows clearly the success of the 'Dutch' pendants which fit with the architecture. (c) The sanctuary. Small pendants are suspended in this area. (d) The saucer domes above the gallery. The concealed downlights can just be seen recessed between each dome. (Architects James Gibbs, Peter Mishcom and Associates for the restoration; lighting consultant Thorn Lighting (Bryan Cross); copyright Thorn Lighting)

(c)

(d)

## BISHOP GROSSETESTE COLLEGE CHAPEL, LINCOLN

The chapel was built at the end of the nineteenth century by Sir Arthur Blomfield in the prevailing neo-Gothic style. An additional north aisle was added in 1901, together with other minor additions in 1932. A major refurbishment carried out by the college architects assisted by their lighting consultants was completed in 1994.

The lighting of the chapel had been judged to be its least satisfactory aspect, both the daylighting and the artificial. The daylighting suffered from sky glare due to the dark surrounds to the windows, while at some period inadequately recessed fluorescent lamps had been applied to the ceiling of the nave, emphasizing irregularities in the dark, poorly painted matchboarding. An early photograph of the chapel illustrating the gas lighting applied at the time clearly shows the darkness of the ceiling and the brick surrounds to the windows.

In considering the upgrading of the lighting, it was agreed that rather than restoring the building to its neo-Gothic style, the aim was to enhance the qualities latent in the original by using currently available materials and techniques for furnishing and lighting. Changes to the interior decoration of the chapel clearly have an effect upon the lighting. For example, the introduction of light-coloured window reveals reduced the contrast between the daylight outside and the interior, reducing glare. Replacing the matchboard ceiling to the nave with white-painted plaster board assisted the reflection of upward artificial light. The method of lighting can never be disassociated from the manner of decoration. The form of the lighting by carefully designed modern pendants, using the latest lamp technology, illustrates the use that pendants can still play in church lighting in both large and small projects.

The main artificial lighting consists of pairs of pendant chandeliers related to the column structure of the chapel. Similar but smaller pendants are employed in the north aisle, and a larger chandelier is

(a)

**Figure 5.8**
Bishop Grosseteste College Chapel. (a) Early photograph of the chapel in 1912 before its refurbishment.

suspended over the balcony. The nine chandeliers in the nave have five arms, supporting a miniature uplight and downlight crowned by five tall artificial candles. While the lamps in the body of the fitting have a closely controlled distribution to eliminate glare, it is the intention that the 'candlelight' is there to be seen and enjoyed for its own sake. The side-aisle fittings have an additional lamp directed to the decorative silver wall panels.

The 7 watt candles, while being electric, are very far from the usual 'electric candle lamp' and have been carefully designed to simulate the glow at the top of a real wax candle, albeit the flame does not flicker. These candles are the same as those developed for Waddesdon Manor by the same consultant.

The pendants are controlled by simple switching to provide the following scenes:

1 'Candlelight' on its own, or together with the following:
2 Upward light to the white ceiling providing reflected light to the interior for general soft illumination
3 Downward light, either on its own or together with one or the other effect providing the main light for reading.

In addition, two spotlights are used to light the reredos which would otherwise be unidentifiable. Additional 'spots' light the notice board and illuminate the stairwell giving access to the balcony.

The flexibility offered by simple switching, whereby pairs of chandeliers are controlled to enable parts of the nave to be lit or left in comparative darkness, enable the chapel to meet the complex needs of the church and the music school. There is a fundamental difference between the soft upward light reflected from the structure for listening and contemplation and the direct downward light required for reading.

**Figure 5.8** (cont.)
(b) Exterior of the chapel in 1995. (Copyright Derek Phillips)

(b)

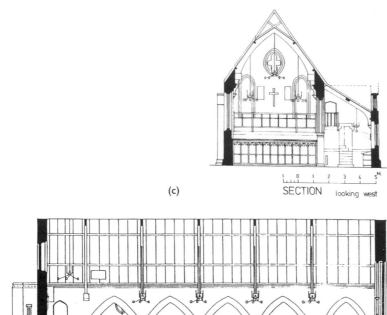

(c)

SECTION  looking west

SECTION  looking north

**Figure 5.8** (cont.)
(c) Plans and sections of the chapel illustrating the lighting scheme. (Drawings by David Medd)

(d)

**Figure 5.8** (cont.)
(d) The chapel by daylight. (Copyright Derek Phillips)

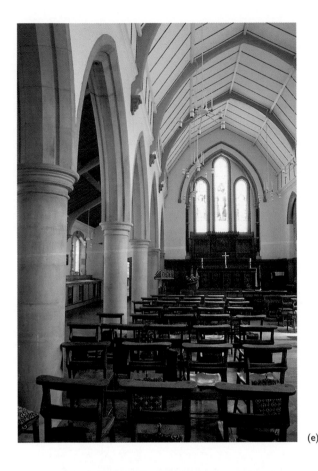

(e)

(f)

(g)

(h)

**Figure 5.8** (cont.)
(e) The chapel by day with added artificial light. (f) The view towards the balcony. Daylight with artificial light added. (g) The aisle. (h) Detail of the pendants with 'candlelight'. (Copyright Derek Phillips.) (Architect David Medd; lighting consultants BAP (William Allen in association with Paul Ruffles))

## ANGLICAN CATHEDRAL, LIVERPOOL

Although strictly a twentieth-century building, the Anglican Cathedral of Christ the King in Liverpool, designed by Sir Giles Gilbert Scott in 1904 is neo-Gothic in design, and fits well into a book on lighting historic buildings. The cathedral is the largest in England, and the architect designed the original pendant filament chandeliers, which, together with the daylight, were to be the main form of lighting. As can be seen in Figure 5.9, the chandeliers still form an important part of the interior, but no longer provide the majority of the night-time light.

When the cathedral was relit in the late twentieth century the brief to the designer was very simple:

1 To reveal fully the interior architecture
2 To accentuate certain features to express the drama of the liturgy
3 To provide sufficient light for the congregation at low level

Other aspects were emphasized, such as efficiency of operation, ease of access to and maintenance of the equipment and flexibility in use. It went without saying that the installation itself should be unobtrusive, with the exception of the existing pendant chandeliers, which were to be retained.

In the words of the lighting designer the object was to 'allow the very vastness of the interior to be seen' by creating a 'flow of light directed across and up the walls'. Modern lighting techniques were used to achieve this, and, as a result, many of the upper parts of the cathedral are now more visible at night than during the day. This is a perfectly valid approach, emphasizing the proposition outlined earlier that the night-time appearance need not be the same as during the day.

The original chandeliers had used 300 watt filament lamps, and these were replaced with 250 watt low-energy sodium lamps. Where filament lamps had been concealed in the interior of the fitting, these were replaced with compact fluorescent lamps, giving more light for less energy and achieving a contrast of colour within the ornate fitting. Use has been made elsewhere of the variety of coloured sources now available, with the main light source being high-pressure sodium, SON designed to reflect the warmth of the structural brickwork. Other cooler sources such as metal halide have been adopted to provide contrast with the overall warmth of the interior.

The aspect of the brief 'to accentuate certain features of the interior' clearly included the altar and reredos, and in a high Anglican church the altar is at the heart of the liturgy, the focal point. The lighting of the altar uses metal halide floods located at high level in the triforium gallery. The result is to bring out the depth shadows and colour and to emphasize the sculptural relief of the figures on the screen.

Although this brief description does not attempt to cover all the intricacies of the lighting and the resulting appearance of the structure, one final point is the use of interior lighting expressed on the outside. At the west end, three high-pressure sodium floods have been incorporated at gallery level, lighting the lancet windows, but also designed to light the stained glass seen from outside.

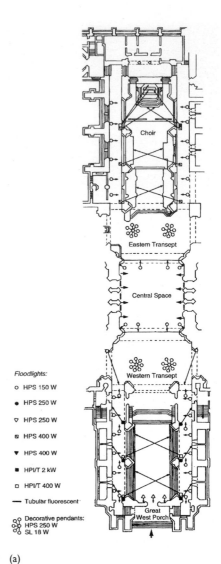

Floodlights:
o   HPS 150 W
●   HPS 250 W
▽   HPS 250 W
◩   HPS 400 W
▼   HPS 400 W
■   HPI/T 2 kW
□   HPI/T 400 W
—   Tubular fluorescent

Decorative pendants:
HPS 250 W
SL 18 W

(a)

**Figure 5.9**
Anglican Cathedral, Liverpool. (a) Plan to illustrate the principles of the lighting scheme.

(b)

**Figure 5.9** (cont.)
(b) View from the west towards the altar. (c) Reredos and screen well modelled by frontal lighting from high level in the triforium. (d) The original pendant 'multilamp' chandeliers modified to accept modern light sources. (e) View towards the altar from the west. (Architect Sir Giles Gilbert Scott; lighting consultants Philips Lighting in association with the Cathedral Architect Keith Scott; copyright *International Lighting Review*)

(c)

(d)

(e)

## STOWE SCHOOL CHAPEL, BUCKINGHAMSHIRE

The chapel at Stowe, by Sir Robert Lorimer, was built in the twentieth century but to traditional designs. Therefore, when it was found that the original lighting from small glass fittings set high in the ceiling was unsatisfactory, a decision was made to improve the lighting. This decision gave the opportunity to meet the modern needs of the school for musical recitals and theatre, an enhanced function.

This was carried out using the lighting techniques available in 1970. The location of the original glass fittings was used to incorporate downlights giving the main light to the nave and congregation. In addition, upward light was used to reveal the beauty of the timber ceiling, from sources concealed at low level in the top of screens. Additional light was added for the altar and the aisles, and for performances special slots were formed in the timber ceiling to accommodate 'theatre spots'. The timber ceiling, one of the attractive features of the chapel, could now be seen and enjoyed without attracting undue emphasis.

Good use was made of the fact that there was ample roof space above the wooden ceiling to permit the maintenance of the downlights recessed into the holes left by the original glass fittings, and to allow for aiming the theatre spots to be used for theatrical performances. Filament sources were employed enabling the whole system to be controlled by a dimmer panel, permitting the priest in charge to vary the lighting according to the needs of the ritual, and this facility has been much used.

**Figure 5.10**
Stowe School Chapel. (a) Plan and section to show the lighting scheme.

(a)

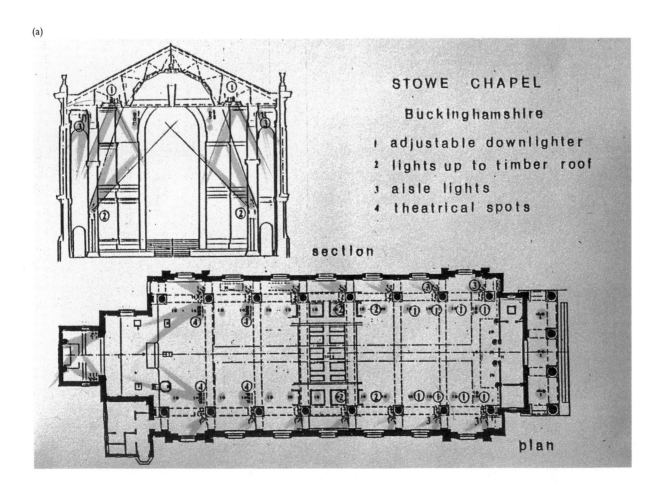

STOWE CHAPEL

Buckinghamshire

1 adjustable downlighter
2 lights up to timber roof
3 aisle lights
4 theatrical spots

section

plan

(b)

(c)

(d)

(e)

**Figure 5.10** (cont.)
(b) The chapel lit by downlight only, with the chancel lights. (c) The ceiling with uplights concealed in the side screens. (d) Detail of the rich wooden ceiling, with uplight, together with the slots for theatre lighting. (e) The chapel with the aisles lit only.

(f)

**Figure 5.10** (cont.)
(f) Detail of the aisle lighting seen from the aisles.

(g)

**Figure 5.10** (cont.)
(g) All the lighting on at full intensity. Dimming allows the different elements to be varied to create a balance suitable for different functions. (Architect Sir Robert Lorimer, Hugh Creighton for the restoration; lighting consultant DPA; copyright Derek Phillips)

## BUDDHIST TEMPLE, DAMBULLA, SRI LANKA

Ecclesiastical buildings are not confined to the Christian religion, and it is of interest to show how the historic buildings of other countries, in this case a Buddhist temple, may benefit from renewed lighting. This seventeenth-century temple is typical of its period, being built under an overhang in the rock with a wall in the front, which virtually eliminates daylight. This temple is famous for its cave paintings, the whole area of ceiling being decorated in this manner. The floor of the cave is laid out with statues of the Buddha, and in the past the only lighting came from clay oil lamps, some permanently lit, in addition to those brought in by the worshippers. Their fumes blackened the ceiling paintings, requiring that they be cleaned at regular intervals. In order to reduce this, in the 1950s the temple authorities had installed a system of fluorescent batten lights, fixed directly to the roof/ceiling to provide sufficient light for the worshippers. This system, while providing light to the temple, was a disaster, causing unacceptable conditions, rendering the paintings difficult to see due to disability glare and destroying the paintings themselves, due to the direct fixing of fittings and wiring to the painted ceiling.

In 1976 UNESCO appointed a consultant to advise the government of Sri Lanka on the relighting of its historic monuments, largely to further the aims of their tourist industry in encouraging visitors to stay longer in the country. The consultant made a number of recommendations, the first of which was to remove entirely the existing system of fluorescent lighting and to repair the damage to the ceiling paintings. Once this was done, a system of indirect lighting would be installed, using as far as possible low-tech sources related to the symbolism of the temple and readily available in the country.

Lighting trials were made using local materials, and these were seen and approved by the minister responsible, but it took some seven years before the money became available to carry out the proposals. Now the ceiling paintings can be enjoyed and the absence of the oil lamps means they no longer get blackened. In addition, the temple has become one of the island's main tourist attractions.

**Figure 5.11**
Buddhist temple, Dambulla. (a) The exterior of the temple in sunlight. (Copyright Derek Phillips)

(a)

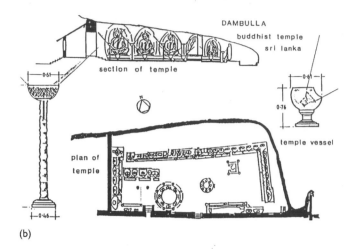

(b)

(c)

(d)

(e)

(f)

**Figure 5.11** (cont.)
(b) Plans and sections showing the use made of temple vessels, using locally available lamps and materials.(c) The temple lighting before it was relit. The fluorescent battens are nailed to the painted ceiling. (d) and (e) Lighting trials in temple. (Copyright Derek Phillips.) (f) A poster featuring the temple after the relighting. (Copyright Jürgen Schreiber.) (Archaeologist Roland Silva; lighting consultant DPA)

## ST MEINRAD ARCHABBEY, INDIANA, USA

While this relighting project is as yet incomplete, and photographs are not available, an explanation of the design method associated with the consultant's sketches is of great interest in illuminating the careful research and sensitive handling of the project. The introduction to the lighting consultants report states as follows:

> We will start the description of the proposed lighting system by imagining we have entered the building late on a sunny summer morning. There are no electric lamps or candles lighted. The sun is warming the north wall, its light tinted as it passes through the stained glass windows. The east, west and south walls are not black, nor is the ceiling or the floor, for they are lighted by the multiple reflections from the first surface being lighted. As the time of day passes and the sun's position changes until it is gone, the evanescence of this daily seasonal event could with study be committed to one's memory. . .the above description is the beginning point of the lighting design. Designed lighting is the intervention by means of electric and candlelight to modify the environment given from the daylight, to serve our needs within the building. We begin the process by adding layers of light, on the various surfaces of the spaces and objects and people, including hymnals, so that every use can be appropriately illuminated to be seen
>
> (Howard Brandston).

The detailed lighting consists of the following equipment:

1 Column-mounted torchères lighting up to the ceiling vaults of the nave
2 Pendant fixtures giving upward light to the aisle vaults, which will themselves gently glow with light
3 Focusable downlights set into the ceiling at the centre of the ornament in the vaults of the nave give the main functional light to the floor of the nave
4 Additional accent and spotlights to be added to highlight certain objects and functions
5 Special candle lamps related to the choir stalls
6 Adjustable fixtures will light the walls of the apse, located in concealed positions behind the columns delineating the apse.

The lighting scheme is well represented by the consultant's freehand sketches, used to present his initial proposals to the client. They show, perhaps better than more carefully drawn later illustrations, the consultant's ideas.

This system provides for lighting all areas of the church, with a control system enabling maximum flexibility for the emphasis of certain areas and the diminution of others as desired by the separate functions of the ritual.

on the one side and recessed panels on the other. The chandeliers were designed to provide light up to the gold ceiling as well as functional light downwards to the table. The architects detailed a 'place' for the wall brackets in the centre of a round panel. The location of each fitting, pendant and wall bracket was therefore predetermined by the decorative design for the interior.

The whole installation was controlled by a microprocessor enabling a variety of 'scene-sets' to be available at the touch of a button, with emphasis on upward or downward light as desired from the chandeliers, the bowl and the arms being on separate circuits. All lamps were on dimmer control, simplified by the use of filament sources.

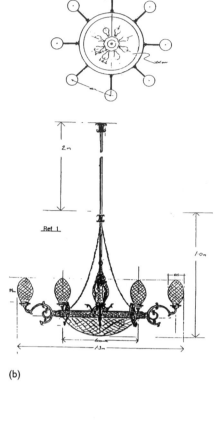

(b)

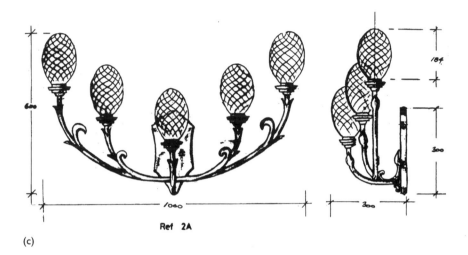

Ref 2A

(c)

(d)

**Figure 6.1** (cont.)
(b) Detailed design for the five pendants. (DPA drawing.) (c) Detailed design for the wall brackets. (DPA drawing.) (d) The completed room by daylight. (Photographer Adam Woolfitt.) (Architect and copyright Cecil Denny Highton; lighting consultant DPA)

### Grand Locarno Room

In the contemporary print for this room illustrating a reception given for the Queen's birthday in the late nineteenth century the light sources are gas jets, and while the light sources needed to be changed to electricity, the layout and disposition of the light fittings, as with the window pattern, provided the basis for the lighting design for the room. The barrel-vaulted ceiling yielded the location for multi-tiered and multi-

**Figure 6.2**
Grand Locarno Room. (a) The completed room. (Photographer Adam Woolfitt.)

(a)

armed pendant chandeliers. Originally there had been three, one to each section of the ceiling. The amount of light given off from the original gas jets would have been minimal, but by increasing the light considerably from the new chandeliers it was considered that the original number of three would be sufficient. The original eight wall brackets were replaced with designs planned to be not dissimilar to those illustrated in the contemporary print.

A pendant fitting was found at Hertford House (the Wallace Collection) which was used as the basis for the castings for the mouldings, as these were contemporary with those at the Foreign Office. It can be seen from the sketch drawn by the consultants for the chandelier manufacturer that the overall design of the chandelier is larger and incorporates a crystal dish at the base, and crystal bowls on each of the three tiers. Upward light is concealed in the central bowls at each level. In addition to this upward light from the chandeliers, fibre optics, using the metal halide lamp, are concealed at high level at the windows lighting up to the vaulting above.

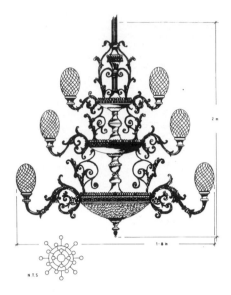

(b)

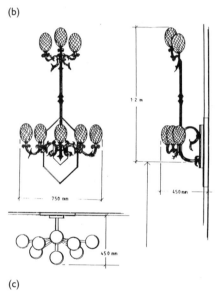

(c)

(d)

**Figure 6.2** (cont.)
(b) Detailed design for the three pendants. (DPA drawing.) (c) Detailed design for the wall brackets. (DPA drawing.) (d) The chandelier at Hertford House used as a model for the castings of the new pendants.

(e)

**Figure 6.2** (cont.)
(e) A contemporary print of the room used as a guide for the design of the wall brackets. (Architect and copyright Cecil Denny Highton; lighting consultant DPA)

## The Foreign Secretary's Office

The daylighting of the room from tall windows on two sides was entirely satisfactory but the artificial lighting installed in the 1950s consisted of a single pendant bowl in the centre of the room, and more light was required at night by the Foreign Secretary of the day, Sir Geoffrey Howe, both on his desk and over the conference table. The central bowl contained a large number of filament lamps, presenting a maintenance problem, but only delivering some 50-lux of light to the desk and even less on the conference table. While being visually acceptable during the day it did little for the appearance of the space in terms of wall surfaces and furnishings after dark.

**Figure 6.3**
Foreign Secretary's Office. (a) The original lighting, condemned as insufficient. (Crown copyright.) (b) The room plans to show locations of pendants. (DPA drawing.) (c) Inspiration for the new pendant found in a lighting manufacturer's old catalogue.

(a)

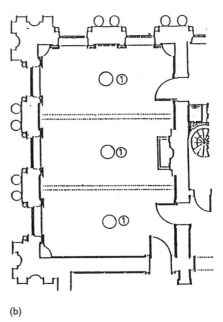

(b)

The criteria for the relighting consisted of a brief to provide a level of light consistent with modern lighting practice (some six times higher than the original) and in a manner appropriate to the design of the room. Since it would have been unthinkable to pierce the decorative ceiling to accommodate forms of downlighting this meant using a form of pendant chandelier. A number of contemporary chandeliers were studied, and the final inspiration for the design came from a fitting made by the manufacturer Dernier and Hamlyn, and illustrated in a catalogue in the 1900s shown at a price of £59.10s. While not being a copy, the chandelier design put forward compares favourably with this original while providing the increased light levels demanded.

The redesigned lighting scheme consisted of three specially designed chandeliers with a central cut-glass bowl concealing fourteen filament lamps, with eight brackets each supporting cut-glass 'pineapples' with filament lamps. To increase the level of light when required, fibre-optic heads were placed in the ring at the base of the fitting fed from a metal halide lamp concealed in the top. This lamp is either on or off, it cannot be controlled by a dimmer system, and suffers from the disadvantage that it requires a 'run-up' period, before coming to full

(c)

brightness. But as all the remaining lamps are filament, these can and are controlled by a scene-set dimmer system, enabling a change of atmosphere to be provided when the room is used for social purposes.

The light to the desk and conference table meets the required levels of light for the functional use of the room, gentle light upwards reveals the beauty of the guilded ceiling, and the wall panelling design and colouring are shown to advantage.

**Figure 6.3** (cont.)
(d) Plan and section: design for the new pendants. (DPA drawing.) (e) Detail of the three new pendants. (Copyright Derek Phillips.) (f) The completed room. (Lighting consultant DPA; Architect Cecil Denny Highton; photographer Adam Woolfitt; crown copyright)

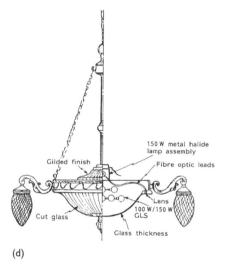

(d)

(e)

(f)

*The Durbar Court*

As a part of the reconstruction of the Foreign and Commonwealth Office in the 1980s the central court of the old India Office was included. The courtyard, designed by Sir Digby Wyatt and completed in 1867, was said to have been inspired by the Italian architect Bramante's cortile in the Palazzo della Cancelleria in Rome, and, as can be seen from contemporary prints, was a fine example of the Italian style. Known as the Durbar Court, it had been used for the entertainment of princes, but had fallen into a sad state of decay and had been unused since the First World War. While it had been open to the sky initially, some sixteen years after its

**Figure 6.4**
Durbar Court. (a) Contemporary print of Durbar Court.

(a)

completion a glass laylight had been added to protect the court from the weather. The new purpose for the court was stated to be for occasional European Community and Foreign Office receptions, but the renovation to the court has meant that it is now used on a regular basis for government entertainment.

As it would be undesirable to attach lighting equipment to the surfaces of the stonework facades, modifications were made to the structure of the roof and laylight to provide catwalks which could be used for mounting and servicing overhead lighting equipment, with the added advantage of facilitating internal glass cleaning. This new roof system allowed a pattern of metal halide lamps to be attached to the rails of the catwalks

(c)

**Figure 6.4** (cont.)
(b) The court in disrepair before reconstruction.
(c) The existing laylight construction to which gangways were added for servicing the overhead lighting.

(b)

lighting towards the side walls, while additional spots were used from the same mounting position to enhance each of the statues at high level.

An important part of the lighting for receptions is the provision of large standard lamps at low level. Eight candelabra were designed, each supporting nineteen glasses containing a single lamp. These decorative fittings, each over 3 metres high, are positioned on each side of the staircases which give access to the court. They are portable and can be moved about to meet the needs of buffets and furniture.

**Figure 6.4** (cont.)
(d) Plan and section of the lighting proposals carried out. (DPA drawing.) (e) Design for the candelabra of which eight were installed. (DPA drawing.) (f) The completed candelabra.

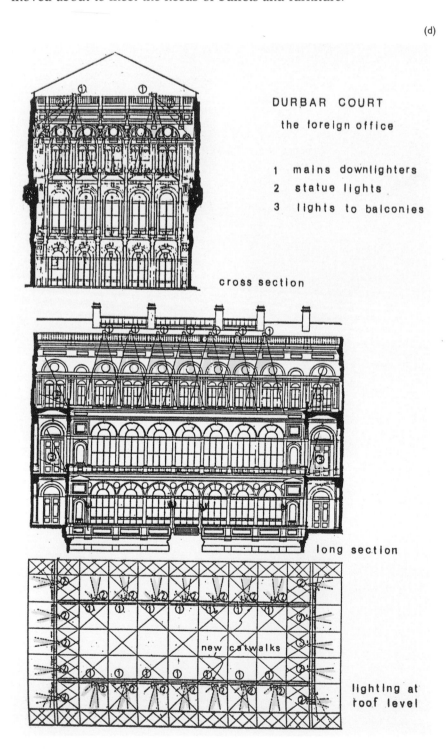

(d)

DURBAR COURT

the foreign office

1  mains downlighters
2  statue lights
3  lights to balconies

cross section

long section

lighting at roof level

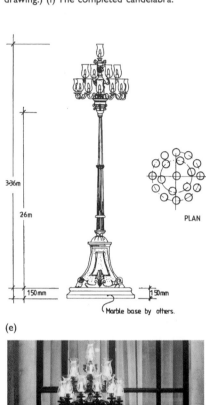

(e)

(f)

There are loggias at each end of the court, and balconies at each level provided concealment for lighting. At the first level the ceiling of the loggia is formed of coloured metal tiles, and these are uplit from below by metal halide lamps, while at the high level the same metal halide fittings give light to the setback of the upper storey.

Finally the interior lighting of the surrounding rooms, seen from the court, makes an impact on the way the court is perceived, and the lighting of surrounding corridors has also received attention.

It would be virtually impossible to build anything along these lines today and it is a tribute to the architects that its quality has been recognized and it has been preserved for future use.

**Figure 6.4** (cont.)
(g) The loggias lit from below. (h) The completed court. (Architects Sir Digby Wyatt (1867) and Cecil Denny Highton (1988); lighting consultant DPA; photographer Property Services Agency; crown copyright)

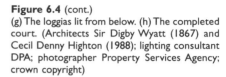

(g)

(h)

## Prague Castle, Czech Republic

Built in the sixteenth century the castle was restyled in the nineteenth century and was the seat of all Czech kings until the Republic was founded in 1918. Since then it has been the seat of the Presidents of the Republic, an historic symbol of Czechoslovak independence. The Rudolf Gallery was built to house the collection of works of art of the Emperor Rudolf II but was subsequently modified in the 1860s for the coronation of Franz Joseph. The chandeliers and wall brackets used at the coronation were originally lit by candles and these have been modified to accept filament lamps, but the richness of detail of the chandeliers has been retained.

Both the Rudolf Gallery and the Spanish Hall are used for state ceremonial occasions, the latter for delegate meetings. The level of light from the chandeliers despite the large number of lamps used (2000 in the Spanish Hall alone) is insufficient to meet modern needs for colour TV cameras, and a backup lighting system has been installed for these occasions. This installation is provided by a permanent mains distribution system which powers a number of fixed positions for 3500 watt metal halide floods. These are placed in position only on occasions when required, but removed at other times to allow the appearance of an interior lit by chandeliers to predominate. The heat from all the chandeliers would clearly be a major problem, and this has been overcome by air conditioning.

(a)

**Figure 6.6**
Prague Castle. (a) The Rudolf Gallery, built in the sixteenth century. (b) The Spanish Hall, built in the seventeenth century. Both halls can be lit to television standards when required. (Architects O. Döbert, Verveka and V. Procházka; lighting consultant L. Morizer; copyright *International Lighting Review*)

(b)

# MUSEUMS

### Royal Museum of Scotland (Main Hall)

Built in 1866, this Victorian building, said to have been modelled on Paxton's Crystal Palace, required great care to be taken with the lighting design. A new roof, an exact replica of the original, had recently been placed in position, and the original bland lighting, from an installation of fluorescent fittings, removed. The roof glazing, rather like the great railway station halls of the Victorian period, gives wholly adequate natural light during the day.

The brief for the night-time lighting design stated that 'The main hall should have all design features enhanced, and revealed without distortion to the proposed colour scheme'. It was implicit in this that whatever lighting equipment was installed should be unobtrusive and not destroy the day-time appearance.

The cast iron columns go from the ground through two gallery levels to support the roof trusses above without a break. As a general rule it is not a good idea to 'break columns in half' by attaching light fittings to them. Therefore, great care was taken to design a fitting which when column mounted at the level of the first balcony would appear to be a part of the structure. Each fitting houses low-voltage lamps together with associated transformers. A medium-beam lamp is directed downwards while a narrow-beam light source provides upward light. This has been the method of lighting up the structure of the roof using low-voltage lamps while others are directed downwards from the same fitting. The size of

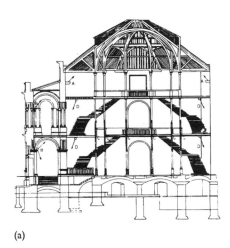

(a)

(b)

**Figure 6.7**
Royal Museum of Scotland. (a) Section showing lighting, through the Main Hall. (b) The Main Hall by day. (Photographer Ken Smith)

the fitting and its colour have been carefully related to the column, and a mock-up was made to convince the client that this would be satisfactory . . . seeing is believing. Additional metal halide uplighting is added to light the opaque areas of roof at the sides from wall-mounted equipment. Low-voltage track is placed at ceiling level in the galleries, from which a flexible system of spots can be directed at the exhibits. Showcases have their own inbuilt lighting.

(c)

**Figure 6.7** (cont.)
(c) The Main Hall at night. (d) Detail across the Main Hall shows gallery lighting. (Photographer Ken Smith.) (e) Metal halide lamp to uplight the roof structure. (Architect The Law and Dunbar Naismith Partnership; lighting consultants Hector Fernandez (Royal Museum of Scotland) and Derek Wilkinson (Thorn Lighting); copyright Thorn Lighting)

(d)

(e)

## Oriental Gallery at the British Museum, London

The British Museum was founded in 1753, but the buildings we enjoy today were built by Robert and Sydney Smirke, starting with the King's Library in 1857. The famous Reading Room, now transferred to the new British Library, was opened during the great Exhibition of 1851, and it is of interest that the trustees decided against the use of gaslight. Visitors had to put up with whatever natural light was available, the museum closing early in winter because of the lack of daylight. Experiments were carried out into the use of electricity, and the British Museum was the first major building in the capital to be lit by electric light as early as 1890.

One room has been selected as being typical of the many different galleries that have been modernized over the years in the Museum; this is the HoTung Gallery, or Oriental Gallery, redesigned and opened to the public in 1992. An early photograph shows the gallery in 1918 before the redesign, lit by carbon arc lamps. The display cases are not

**Figure 6.8**
Oriental Gallery, British Museum. (a) The Gallery in 1918 with carbon arc electrical light fittings. (Copyright British Museum.)

well lit by daylight, and indeed many of the displays would have been difficult to enjoy, placed as they were at low level in closed showcases.

The daylighting received from the tall windows along the north face of the building towards Montague Street has always been excellent and this has been exploited in the more open displays of the new Gallery. Display lighting has now been incorporated at low level in the new showcases, and overhead miniature spotlights are directed on to the taller displays from track-mounted fittings concealed at high level related to the ceiling beams. No general light is provided, as the displays themselves are the objects of attention, and sufficient light is reflected off these for visitors to negotiate the gallery in safety.

**Figure 6.8** (cont.)
(b) The Gallery as refurbished: daylight. (Copyright Derek Phillips.) (c) The Gallery as refurbished: night. (Copyright British Museum.) (Architects and lighting consultants British Museum staff)

(b)

(c)

## Imperial War Museum, London

The Imperial War Museum is an example of a 1980s refurbishment of an historic Victorian museum, where the architects created a large amount of extra space by glazing over the central courtyard. It is this tall daylit area, with its unique artificial lighting solution, rather than the lower gallery spaces that forms the subject of this study.

The high levels of natural light entering through the overhead glazing presented a problem of contrasts, as between the central area during day filled with natural light and the lower surrounding gallery spaces. During the day the central area can be very bright, and the side galleries have accordingly to be highly lit to avoid them appearing gloomy. At night the reverse is the case, and, albeit at a lower ambient level, the galleries appear bright with a more sombre central area. The problem was solved by flexible electronic control, carried out by photocells linked into a time clock. These control the dimming and switching automatically to eliminate human error in the management of the lighting system.

The method of lighting adopted was 'theatrical', using low-voltage spotlights suspended from theatre 'lighting bars' at high level directed down to illuminate the exhibits. These high-powered spots, having a 'throw' of some 30 metres, can be adjusted to reflect changes in the layout of the exhibits, some of which are large aeroplanes hung from the roof level. At night the central roof area is flooded with filtered 'blue' light to relieve the dark sky effect that would otherwise be created, providing an effective background to the undersurface of the suspended planes, which are themselves lit from below by specially adapted Second World War searchlights at ground level. These searchlights throw their light upwards from concealed tungsten halogen and metal halide sources, and, as in their traditional role, can be directed towards their specific 'target'. The whole system is flexible to meet the needs of changing exhibits, the time of day and the weather outside the building.

**Figure 6.9**
Imperial War Museum. The central covered courtyard (a) by night and (b) by day. (Architect Ove Arup Associates; lighting consultants Lighting Design Partnership in association with Jasper Jacob Associates; copyright Lighting Design Partnership)

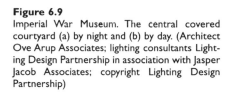

(b)

(a)

## EDUCATION

### Royal College of Physicians, Edinburgh

The College buildings were designed by the architect Thomas Hamilton, in the early nineteenth century. His original intention had been to create 'a temple-like building of considerable grandeur', and this he did with ornate interiors, splendid ceilings, friezes and marble columns. The main accommodation benefits from daylight. The design intention for the refurbishing in 1994 was to retain the historic lighting as far as possible, but to upgrade this by enhancing certain architectural details while increasing the level of light to meet modern standards.

In the Great Hall the large central chandelier and those in the arcades were renovated and the lamps changed to smaller, long-life filament lamps, and the cove at high level was used to house tungsten halogen sources to light up the frieze with its statues. These provide added light to the ornate

**Figure 6.10**
Royal College of Physicians. (a) The Great Hall.

ceiling, but the increase in light level is achieved by the incorporation of warm metal halide lamps mounted outside the windows in the curved area of ceiling. The whole is controlled by automatic microprocessor 'scene-sets' enabling the lighting to be balanced to overcome the preponderance of light at high level.

The library is uplit using fresnel spotlights recessed into the tops of the bookcases, boosting the light available from the original chandeliers by reflection from the coffered ceiling. Four of the existing bookcases were converted into display cabinets for rare books, lit using fibre optics to retain the low temperature required.

It had been the architect's intention that the Entrance Hall and Grand Staircase should have the appearance of an 'exterior' courtyard. To implement this the tall vaulted ceiling was lit by blue 'cold cathode' lamps from the surrounding cove, and discharge sources were again used to light through the artificial 'windows' at high level, this provides the functional light. In addition, busts and niches were enhanced by concealed low-voltage spots, and picture lighting was provided for two large portraits.

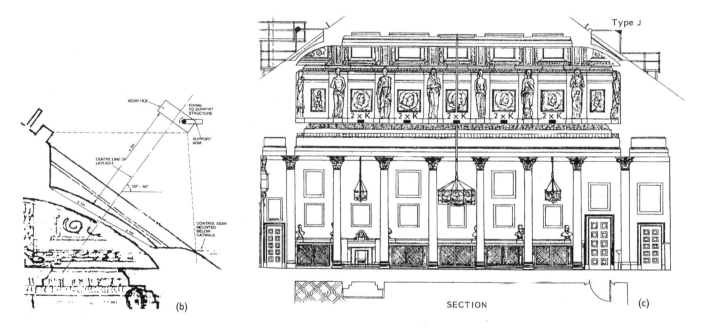

Figure 6.10 (cont.)
(b) Section of the window lighting in the Great Hall. (LDP drawing.) (c) Section through the Great Hall. (LDP drawing.) (d) The Library.

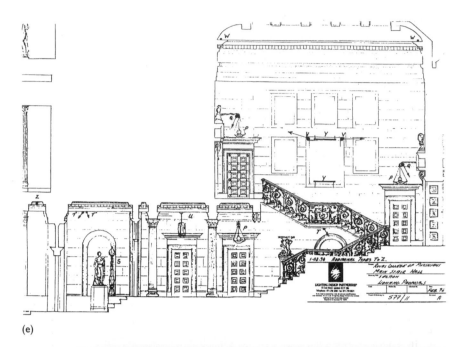

(e)

(f)

**Figure 6.10** (cont.)
(e) Section showing the Entrance Hall. (LDP drawing.) (f) The Entrance Hall. (Architect for the refurbishment Benjamin Tindall; lighting consultant and copyright Lighting Design Partnership)

## Whitworth Hall, Manchester

This is one of the buildings of Manchester University in the late Victorian Gothic style designed in 1903 by the architect Paul Waterhouse. The Hall is used for public meetings, concerts, examinations, lectures, degree ceremonies and social functions. For this reason, the lighting design was required to provide a wide set of parameters, to meet the contemporary needs of the space. . .an 'enhanced' use.

It was clear that the lighting should reflect the 'High Gothic' interior with its great hammerbeam roof and enhance the large organ from the decorative aspect. In addition, it had to provide a high level of light to the interior on occasions to meet the visual needs of those using the hall.

When the hall was built it had been lit by gasoliers suspended from the hammerbeams, although these had been replaced in the 1930s by large hexagonal pendants, described as 'vaguely Gothic in shape', containing filament lamps behind white opal glass panels. With these being the only lighting to the hall, the general effect left the huge organ case and the fine roof in gloom, and the general effect was heavy and depressing. The level of light was so low that when the room was to be used for examinations it had been necessary to suspend a series of fluorescent batten fittings from the tie bars above.

Because of the wide range of lighting requirements, it was decided to meet these by a floodlighting scheme dispensing with the pendants as the major source of light. Fortunately, a number of late nineteenth-century wrought iron 'gaslight' fittings were located in a redundant church in Bolton, and these were converted to electricity to replace the existing 1930s pendants to form the main decorative light for the hall. In refurbishing these pendants, low-voltage lamps were used to replace the original gas jets (underrun to extend their life) and a floodlamp mounted in the base directed upwards to illuminate the ornamental gilding. Despite contributing emphatically to the visual quality of the Hall (these are what the visitors judge to be its main lighting) the chandeliers on their own are little more than decorative objects creating sparkle and atmosphere. The real work, to satisfy the new needs of the space, is performed by the functional light fittings placed at high level. Floods of 1000 watts, directed downward from the flat ceiling at the apex of the roof, light the central area, and are so high and concealed by the beam structure that they do not constitute a glare source. The roof is lit by 500 watt floods directed upwards from the level of the column tops onto the slope of the roof and upper areas of the walls. The middle parts of the walls are lit from fittings mounted at the ends of the hammerbeams, with spill rings to control glare. The lower walls and paintings are lit from fluorescent lamps concealed behind baffles painted to match the wall colour, giving both downward and upward light. The organ case and platform are illuminated by stage lighting techniques from carefully sited vertically mounted fittings.

The result of the various lighting systems installed ensures that, although the visual impression is of the decorative pendants, the lighting now provides, by means of controlled switching, the high or low levels of illumination needed to meet the multiple use of the space. The architectural form and detail of this splendid interior have been enhanced, while the lighting satisfies its multiple function.

**Figure 6.11**
Whitworth Hall. The view towards the organ.
(Architect Paul Waterhouse; lighting consultants
Thorn Lighting in association with James Bell and
D. Buttress; copyright Thorn Lighting)

## Marlborough College Library, Wiltshire

The school library, a fine example of nineteenth-century 'public school Gothic', was lit by pendant fluorescent fittings. While the amount of light was satisfactory and illuminated the shelving adequately, the fittings themselves were glaring, obtrusive and out of sympathy with the architecture, causing a visual conflict with the roof structure.

The following simple proposals were put forward: first, to adapt the top of the book stacks to allow lamps to be concealed, lighting up to reveal the quality of the roof construction and, second, to replace the fluorescent fittings with a less obtrusive but still functional fitting. The new installation was carried out in 1992.

The lamp chosen for the upward lighting was the compact fluorescent, giving efficient upward light of a colour which would enhance the roof structure, three lamps being installed in the top of each book stack. The fitting chosen to replace the fluorescents was a decorative prismatic glass pendant containing the 250 watt tungsten halogen lamp, twelve in all being employed. While this increased the electrical load, glare was eliminated and it was now possible to use the library in its functional sense, and the improvement in the environmental quality of the library was significant.

**Figure 6.12**
Marlborough College Library. (a) The Library before relighting with the fluorescent installation.

(a)

(b)

(c)

(d)

**Figure 6.12** (cont.)
(b) The Library lit with uplight only. (c) The uplit roof construction. (d) The completed lighting installation. (Architects Bodley and Garner; lighting consultant DPA; copyright Derek Phillips)

## LEISURE

### The Landmark Hotel (formerly the Regent), London

Designed originally as the Great Central Hotel by Robert Edis at Marylebone Station, this opened in 1899 and was considered at the time to be one of London's best and most luxurious Victorian 'railway terminus' hotels. Only 40 years later it closed its doors. The hotel had had a chequered life, being requisitioned during the First World War as a convalescent home for officers, in common with many other properties in London. In the Second World War it served as a transit station for troops and accommodation for soldiers on leave. At the end of the war, with inevitable dilapidation, it was converted into offices for what is now British Rail, serving as its headquarters until 1986. The present use takes it back to its original 'luxury hotel' status, and it

(a)

**Figure 6.13**
The Landmark Hotel. (a) The completed Winter Garden in the original open courtyard, eight storeys high. (Photographer Robert Miller)

**Figure 6.13** (cont.)
(b) Central courtyard reconstruction. (Photographer Robert Miller)

opened as the Regent in 1993, after nearly four years of restoration. In August 1995 it was renamed the Landmark Hotel.

The most dramatic area of the hotel is the central atrium, originally an open courtyard, 100 ft by 125 ft, serving as access for horse-drawn carriages, but, in addition, providing daylight to all hotel guest rooms. This area has been covered as a winter garden, and is the prime element of public accommodation, both by daylight and at night. The night-time lighting is derived from the activity at ground level, allowing spill light onto the surrounding facades, and a feeling of moonlight above, while light spills from the surrounding windows of the eight-storey court.

Many of the original public spaces have been refurbished in the Victorian style, an example of which is the drawing room, seen in 1899 in Figure 6.13 (c), lit with delicate pendant gasoliers, while the same room in 1993 is lit in a similar manner, with fewer larger multibracket electric chandeliers.

**Figure 6.13** (cont.)
(c) The original drawing room at the turn of the century. (d) The drawing room today. (Photographer Robert Miller.) (Architects Robert William Edis (1897) and Geoff Reid and Associates (1990); lighting consultant Lighting Design Ltd (Sally Storey); copyright Landmark Hotel)

(c)

(d)

## The Reform Club, Pall Mall, London

The Reform Club, designed by Sir Charles Barry in 1837, was inspired by the Farnese Palace in Rome 300 years earlier, and is one of the finest of London's political clubs of the Victorian period. The day-lighting in the club was excellent, but the artificial lighting was in need of upgrading and the brief to the consultants in 1985 consisted of a list of priorities, the most important of which emphasized the need to retain the present character, while increasing the light levels in places. Where possible, to reduce the electrical load and improve access for maintenance, the work was confined to the central atrium area and surroundings.

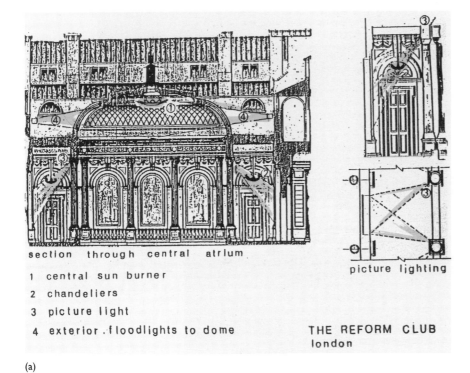

section through central atrium

1  central sun burner
2  chandeliers
3  picture light
4  exterior floodlights to dome

picture lighting

THE REFORM CLUB
london

(a)

**Figure 6.14**
The Reform Club. (a) Section through the atrium (DPA drawing.) (b) 'Sunburner' fitting lowered for maintenance. This shows the original filament lamp holders for 300 watt GLS lamps: eight lamps use 2400 watts. (Copyright Derek Phillips)

Three specific points were made: the lighting of paintings should be improved, the central 'sun burner' fitting retained and the indirect fluorescent lighting to the coffers below the first-floor balcony either improved or removed. The central atrium was lit by daylight through a glazed and faceted rooflight, which gives gentle daylight to the whole of the central area, fine during the day but is less satisfactory when dark at night. A decision was made to cast light through the glazed facets from outside the roof at night, and this was achieved by the use of eight high-pressure sodium floods placed two on each outside wall, shining down into the atrium. This ensures that the atrium ceiling gives a warm glow at night.

The 'sun burner' fitting, which would originally have been powered by gaslight, was first lowered for inspection by its raising and lowering gear, and it was found that of the nine filament lamps, only four were burning. The life of these lamps is 1000 hours, and less when heated in a confined

(b)

## Prince Edward Theatre, London

The theatre was designed by E. A. Stone in the 1930s in line with the Art Deco interiors of the period, and after a number of changes of use – from cabaret restaurant, to cinema and back to a theatre in 1974 – it was in great need of refurbishment. The architects were asked to refurbish the theatre and to investigate the role that lighting might play in improving both the interior and exterior appearance. The outside appearance is covered in Chapter 8. The entrance foyer has been returned to its original circular shape, with a central artificial laylight lit from above using cold cathode lamps. This is surrounded by a ring of low-brightness downlighters, to give unobtrusive functional light to the space. The perimeter is defined by a glazed decorative pelmet backlit using low-voltage xenon festoon lights and column heads are internally illuminated. The lighting is circuited to enable a higher level of light during the day, when the light must 'compete with daylight,' and lower at night.

The architect solved the complex planning problems presented by the needs of a modern theatre, but returned to an Edwardian concept, with a series of new boxes down each side. These and other changes permitted considerable improvements to be made to the lighting. The auditorium now recreates the 1930s atmosphere, using the original decorative sources, the illuminated fascia lights to the front of balconies and the large-scale wall lights, but using up-to-date technology. The wall lights, consisting of a series of faceted glass tiers, have been modified to accept low-voltage capsule lamps and cold cathode rings. To back up this 'decorative' light, low-voltage downlights are concealed in the soffits of the upper and dress circles, while powerful down-lighting was added to light the front stalls from a considerable height.

The concept was to separate the decorative from the functional light, so that while the appearance gives an authentic Art Deco appearance, the levels of lighting meet modern standards. All lighting, including

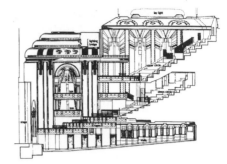

**Figure 6.15**
Prince Edward Theatre. (a) Section through building.

**Figure 6.15** (cont.)
(b) The circular foyer.

the cold cathode sources, are dimmer controlled to achieve flexibility and to emphasize the functional change of atmosphere from the high levels while the audience take their seats, to decorative lighting once the show is about to start, and finally the stage lighting only, a new lighting bridge being installed to deal with the last. The architects, together with the lighting consultants, have created a modern facility, having all the qualities associated with the best decoration of its original period.

(c)

(d)

(e)

**Figure 6.15** (cont.)
(c) Detail of tiered wall lights. (d) Wall light. (e) The Auditorium. (Architect RHWL Partnership, lighting consultant Lighting Design Partnership; photographer RHWL/John Walsam; copyright Lighting Design Partnership)

# TRANSPORT

## Euston Station, London

Where the great train station buildings of the nineteenth century have survived, the structure has for the most part been allowed to remain unaltered, necessary changes being made at ground level within the span of the arches above. Euston is one of the exceptions, where the beautiful curved Victorian roof structure was replaced by a low-level 'utility' roof. Both serve the functional purpose of covering the platforms and admit daylight, the original provided it with 'delight'.

Many of London's Victorian stations survive today, such as St Pancras, King's Cross and Paddington, and York in the north. Modifications at low level have clearly been necessary to meet modern requirements, but the overall roof coverings serve the same purpose today as yesterday. There has been little need to change them, the daylighting was well thought out, and the industrial night-time lighting applied, changing with developments in lamp technology, appears quite appropriate.

(a)

(b)

(c)

(d)

**Figure 6.16**
(a) The original trainshed at Euston station before modernization. (b) The platforms after modernization. (c) York station's trainshed in 1995. (d) King's Cross station's trainshed in 1995. Both are comparatively unchanged. (Copyright Derek Phillips)

## Liverpool Street Station, London

In contrast to the redevelopment of Euston, and arguably because of it, the brief to the architects insisted that the original Victorian trainshed be retained and that the station be treated with respect as an historic building, thus ensuring that the quality of the architectural and engineering design was preserved, and, if necessary, extended. At the same time, it was a requirement to create an efficient new railway terminus to meet the needs of the twenty-first century.

The original station, where work had started in 1875, was in the great tradition of Victorian railway termini built in the capital in the late nineteenth century. The reconstruction of Liverpool Street Station carried out in 1991 retains not only the Victorian trainshed 'daylit roof,' over the platforms but continues this with similar construction over the main concourse.

Considerable work was required to preserve and, where necessary, to repair the original Victorian Gothic cast iron columns and roof over the platforms. The architects used similar details to engineer the new roof over the concourse in such a manner that it is difficult to discern where the old stops and the new begins. The architects have managed to marry the two, without in any way prejudicing the needs of the terminus.

Liverpool Street Station is really a 'daylighting' story. The artificial lighting essentially follows the industrial lighting methods adopted in the past but uses more energy-efficient light sources and equipment, with fittings placed at high level below the glass roof with exterior access for maintenance.

**Figure 6.17**
Liverpool Street station. The trainshed (a) by day and (b) by night. (c) The concourse by day viewed towards the information board. (d) The concourse by night viewed away from the information board.

(a)    (b)

(c)    (d)

(e)

(f)

**Figure 6.17** (cont.)
(e) The new shops below which access is gained to the platforms. (f) The new entrance, designed to fit well with the modernized station. (Architect for reconstruction Nick Derbyshire Design Associates Ltd; copyright Derek Phillips)

## Baker Street Underground Station, London

The Victorian station at Baker Street, which formed a part of one of the world's first underground railway systems, was originally lit by daylight coming through shafts above the platform walls. These shafts served a double purpose, allowing smoke and steam to escape, but after some years they became redundant, due to alterations at street level above, and were boarded up below behind advertising hoardings.

As part of the London Transport's station improvement programme, the hoardings were removed to reveal the original yellow London stock bricks, and high-pressure sodium light sources were concealed in the embrasures to duplicate the original daylight. The result, which concentrates light where it is required on the platforms, replaces a system of fluorescent light above the platforms and is of particular interest as an integration of light and structure, albeit originally for natural sources.

**Figure 6.18**
Baker Street Underground station. The original filled-in daylight opening is used again to house artificial lighting employing the high-pressure sodium (SON) lamp. (Lighting consultant Thorn Lighting; photographer Alan Turner; copyright Thorn Lighting)

## CONCLUSIONS

Obviously, in any refurbishment of the artificial lighting of an historic building the method of daylighting is crucial during the day, and nothing should be done to interfere with this. However, the artificial lighting may need to be designed to supplement daylight during the day, where it is inadequate for the function of the building.

The above examples, while not in any way exhaustive, define a number of the approaches to the relighting of public buildings. There is always room for innovative approaches to the problem and new lighting technology will no doubt expand and increase the possibilities open to future designers. However, the following conclusions can be drawn from these examples:

1 Where the lighting needs to be 'improved' this does not always mean that an increase in light level can be justified. It may mean that the manner of the original lighting is glaring or inappropriate in other ways.
2 An analysis must be made of the function of the room, which may need the lighting to be enhanced to meet new requirements. The most important criterion is that the new lighting must satisfy any change to the function of the space.
3 The lighting must be appropriate to the interior decoration of the space. This may be achieved by the adaptation of the existing lighting fittings to meet modern needs, or by the addition of light fittings in an unobtrusive manner. This may suggest incorporation of recessed lighting in ceilings, greatly assisted by miniature lamp technology, but always being careful to avoid the visual destruction of decorative ceilings.
4 Glare is to be avoided at all times. This is not just the elimination of glare at normal angles of vision but also glare which emanates from bright fittings at high level, making it difficult to appreciate the architecture, often associated with the lighting of tall structures such as cathedrals. There is, however, a difference between 'glare' and 'sparkle'. Whereas a candlelit chandelier may be seen to produce glare if taken on its own, it may be entirely appropriate when seen in its context as a part of the visible contrasts in a space.
5 There is a need to investigate the most up-to-date methods of lighting technology to see if there are new means to achieve the decorative or functional criteria while retaining the desired decorative appearance.
6 There is a requirement to cooperate with the structural engineer, in addition to the architect, to ensure that whatever changes are made are safe and can be maintained easily. This may mean some modifications to structure.
7 Methods of 'uplighting' may be at the heart of new proposals. The easy option of incorporating light sources in the top of cornices at high level may seem attractive but suffers from the 'high brightness' associated with light reflected from white ceilings, and should be used sparingly. It should never be left to the ceiling to provide the only functional light at low level. Where it is important that the ceiling is lit for decorative reasons, there will generally be methods, other than cornice lighting, that can be adopted, such as sources incorporated into furniture units, portable fittings or concealed wall lighting, or indeed from suspended fittings.

8 Finally, the means now available for the control of lighting has added another dimension to the possibilities open to the designer, whether by simple dimmers or by programmable microprocessor systems. The control systems now available can reflect the changing seasons of the year, the times of day, or the function of the space, adding immeasurably to the flexibility of the lighting solution.

# 7 Change of use

When analysing the problem of an historic building for which an entirely new use is planned, several aspects are important. A decision has been made that the original use of the building no longer exists, and therefore what remains of the building is a vehicle for reconstruction for its new use. It must not be a straitjacket, but it will impose its own discipline in a number of ways.

We are not considering the historic building which has outlived its usefulness, and where the decision has been made to pull it down and put up a new one. In many cases there will be arguments for this approach, not least that it may be a cheaper solution. Our concern is with the national heritage of buildings for which the imperative is for continued utilization, rather than demolition or sterile preservation.

The extent of the reconstruction will depend upon how closely the original building corresponds to the functional requirements of the new. For example, a stately home may be altered to a particular style of hotel without serious change to the character of the interior, while a redundant hospital would require substantial alterations to achieve the same effect. A derelict brewery might be converted into an art gallery with some difficulty, while a Victorian trainshed might retain its original daylit appearance, with the necessary alterations to convert it into an art gallery occurring only at low level.

As always with any modernization or reconstruction of an historic building, the important aspect of daylight must be addressed. Daylight affects the exterior appearance of the building as well as the interior, and in most cases will impose its own discipline in terms of window patterns to external elevations. It may be possible to provide for overhead daylight by glazing over otherwise solid areas to allow natural light to permeate some internal spaces, but in all cases the spatial possibilities of the building will derive initially from daylight. On the other hand, the artificial lighting systems employed should be state-of-the-art technology. Basically, 'daylighting is for ever' but 'artificial lighting is constantly developing'. As it relates to the interior design, lighting may need to change for good technological reasons or in some architectural programmes change to meet the needs of fashion.

The key to success in this field is one of maximizing the opportunities available to express the qualities of the original historic building while satisfying the functional needs of its new use. A number of examples follow which demonstrate how the intrinsic character of the original building can be retained.

## ARCHITECT'S OFFICE, PRESTON

A nineteenth-century industrial mill building in Preston which was converted first to a biscuit factory in the early twentieth century and later to a design office by an architect for his own use demonstrates the flexibility of such buildings. None of the original interior was retained, but the newly created space, reflecting the needs of modern servicing in lighting and acoustics in the 1960s, retains the original window pattern, exposed brickwork and cast iron columns, expressing its origins.

(a)

(b)

**Figure 7.1**
Architect's offices. (a) The original Powell's biscuit factory. (Photographer Building Design Partnership.) (b) The architect's offices created from the old factory. (Photographer James Bell.) (Architect, lighting consultant and copyright Building Design Partnership)

# THE BUSINESS DESIGN CENTRE (BDC), ISLINGTON, LONDON

The BDC, originally the Royal Agricultural Hall and a well-known London landmark, was completed in 1862, and was described by a contemporary as 'well entitled to a place among the prominent features or institutions of the metropolis, it is a gigantic building, of the railway station order'. The architect was Frederick Peck, but the building was the brainchild of the Smithfield Club, founded in 1778, to promote improvements in agriculture. Contemporary prints show its opening exhibition, lined with cattle stalls, and for many years there were annual exhibitions, some of which were attended by royalty. In 1880 the first of many military tournaments was held, and the hall was used for different activities, including Cruft's dog shows, which continued there until 1939. However, with the outbreak of war in 1939, the 'Aggie', as it was affectionately known, was commandeered by the government and when a fire destroyed the postal sorting office at Mount Pleasant the operation was transferred and remained so until 1970. By 1981 the building had been deserted for eleven years, and had become so dilapidated that there were plans to pull it down for redevelopment. The BDC, as it now is, had been placed on the draft statutory list of historic buildings in 1971, meaning that it would be difficult to obtain permission to demolish it, and this, together with the foresight of the owners, sowed the seeds of its present use as a successful exhibition hall and trade centre.

**Figure 7.2**
The Business Design Centre. (a) The original market in a state of disrepair in 1980. (b) The reconstructed building in 1995, with the 'Country Living' exhibition, during the day.

(a)

(b)

The daylighting of the exhibition hall remains much the same as originally, with glazing introduced into the great span of the shed roof. The artificial lighting derives principally from the nature of the individual exhibition in progress in the main hall, with the specialist lighting reflecting the needs of the individual showrooms at gallery level. This gives light where it is needed. In addition, uplighting is added from sources placed above the gallery level directed upwards to 'lighten' the curved roof.

(c)

(d)

**Figure 7.2** (cont.)
(c) The roof daylighting. (d) Artificial lighting to the lower roof area. (e) An exhibition held during the Hilight Conference in 1996 (seen at night). (Architect RWHL for the restoration; lighting consultant DPA; copyright Derek Phillips)

(e)

## CATHEDRAL MUSEUM, FULDA, NORTH GERMANY

This is a restoration and conversion of a neo-gothic wing of the seminary and of a Baroque deanery adjacent to the cathedral founded by St Sturmius in AD 744. A new block was inserted between these. The conversion is the work of a British architect, who also designed the lighting scheme, collaborating with a local firm of architects in north Germany.

The main space of the seminary chapel is vaulted, and showcases are placed at an angle along the line of the vaults, from the top of which concealed indirect fluorescent lighting illuminates the vaulting, and is the only general light to the space. Other light from low-voltage adjustable fittings recessed into the sides of the showcases lights sculpture exhibits placed along the walls. The showcases themselves are internally lit, allowing the main historic structure to remain free of any hardware which might interfere with the simple geometry of the space.

The Chapel of the Deanery is now an exhibition space for a silver and gold altar. This is placed within a very large showcase, allowing the altar to be removed on feast days, and is lit by lighting track placed vertically on either side. General lighting is increased by free-standing uplighters. An old cupboard previously in the sacristy has been glazed and now acts as a showcase in the old Deanery with lighting carefully concealed to light up the exhibits within.

The results of this conversion demonstrate the advantages of close cooperation between the architect and the lighting consultant (in this case the architect is a lighting consultant). It demonstrates how the aesthetic qualities of the original building assist in satisfying the functional requirements for its new purpose.

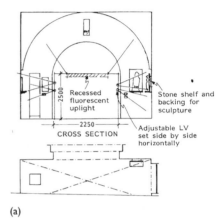

(a)

**Figure 7.3**
Cathedral Museum, Fulda. (a) Section illustrating the lighting design. (b) Plan indicating the showcases and vaulting.

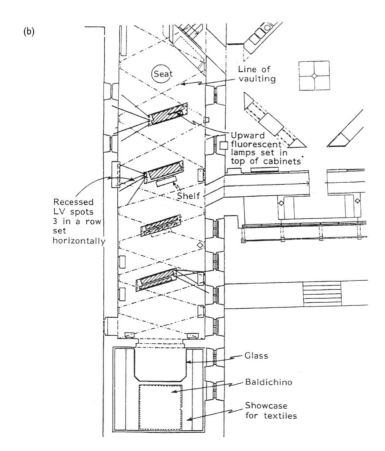

(b)

(c)

(d)

(e)

(f)

**Figure 7.3** (cont.)
(c) General view of the Seminary Chapel. (d) Gallery for medieval art, with overhead fluorescent picture lighting. (e) Chapel in the Deanery. (f) Showcase in the old Deanery. (Architect Michael Brawne in association with Ollertz and Ollertz; lighting consultant and copyright Michael Brawne)

## SAN ANTONIO MUSEUM OF ART, TEXAS

The conversion of a nineteenth-century brewery which had remained derelict for some years is an example of a building which had outlived its usefulness and had been threatened with demolition, but where it was felt that its continued existence as an important feature of the town merited its reconstruction for a new civic use, that of an art gallery. Built on three levels, the original Lone Star Brewery is supported by cast-iron columns with a contoured floor slab to the first two floors, while daylight enters above through a new sawtooth rooflight. A daylight study was carried out, and the original windows retained with some modification and some additional openings to ensure adequate daylighting for its new function, albeit controlled by grey tinted glass and external blinds to assist the needs of conservation. The lighting to the lower floors left the original structure intact by adding track-mounted spotlighting below the contoured floor slabs. These direct light away from the ceiling surface, concentrating it where it is required on the works of art, and gaining the flexibility offered by a track-mounted system, allowing changes to be made to suit the changes in display.

At the upper level artificial sources were added below the rooflight to complement daylight, the lights being concealed behind deep coloured baffles at right angles to the sawtooth roof. The solution provides a modern gallery facility, retaining the original historic quality of the old brewery. From the outside, the new museum of art carries on the architectural tradition and street frontage of San Antonio.

**Figure 7.4**
San Antonio Museum of Art. (a) The Lone Star Brewery before restoration. (b) The original area with contoured floor slab before restoration. (c) Gallery completed below the sawtooth roof. (d) Gallery completed in the contoured floor slab area. (Architect Cambridge Seven Associates Inc.; lighting consultant H. M. Brandston and Partners Inc.; copyright Nick Wheeler)

(a) (b)

(c) (d)

## CLIVEDEN, BUCKINGHAMSHIRE

Cliveden, owned by the National Trust, was built in 1851 by Sir Charles Barry, and was the home of the Astor family, of whom Lady Nancy Astor was well known as the first woman member of parliament. The house and grounds have been let as an hotel, and while this is clearly a change of use, at the same time the buildings do not require any major reconstruction to adapt them to meet the functional requirements of this particular hotel: an hotel designed to provide its guests with the authentic flavour of the original stately home.

The majority of the refurbishing work was in fact concerned with the services to the hotel to make the guests comfortable, since the stately homes of the past were notoriously ill provided in this respect. Hotel guests nowadays, while wanting all the appearance of a more spacious age, demand that they are warm, that there is plenty of hot water, and that the kitchens have been modernized. These necessary improvements are not immediately perceived by the visitor. It is the atmosphere of the reception areas which is of more direct concern, and this is where in this instance the problem becomes more akin to that of enhanced use rather than a change of use.

(a)

(b)

The long drawing room used as the entrance foyer and lobby is a fine example of a mid-Victorian interior. It is both welcoming and comfortable, and with well-considered daylighting coupled with large table lamps, little change was needed. The French dining room remains unchanged, and is ideally suited to its purpose. The main hotel dining room, on the other hand, was converted from the redundant library and indicates how it is possible to get things wrong. This book sets out specifically to consider only successful projects. . .it would be too easy to illustrate mistakes. The library at Cliveden is an exception, and only because it has been corrected, and makes a good object lesson of how not to treat a fine room. Fortunately, an illustration exists of the original library, with which the first reconstruction can be compared. It can be seen that the room had a timber ceiling, with timber panelling between the bookshelves. Originally the room had excellent daylighting and the artificial lighting consisted of light reflected from the picture lights to the paintings, coupled with table lamps.

**Figure 7.5**
Cliveden. (a) Exterior view from the entrance towards the house. (b) The hotel foyer/lounge. (Copyright Derek Phillips)

(c)

In converting the library to a hotel dining room the timber ceiling was removed, and a plain reflective plaster surface installed, allowing upward light from fluorescent lamps concealed in the surrounding cornice to provide uniform light to the room. This was the main change to the artificial lighting, but associated with this, the timber panelling to the walls and bookshelves had been covered by cladding and painted white, thus eliminating the original picture lighting. The new room design, of which the lighting played an important part, was out of character with the rest of the hotel and for this reason would have been unacceptable to the guests, evidenced by the changes made soon after it had been done. The room has now been restored, and is still used as the hotel dining room. The wall cladding has been removed to reveal the original timber panelling and bookshelves and the fluorescent cornice lighting has been replaced by two large crystal chandeliers. During the day, these chandeliers, associated with the available daylighting, provide acceptable conditions. At night the chandeliers are dimmed, thus allowing fibre optic lenses concealed in the top supports to the chandeliers to illuminate the paintings.

While the hotel at Cliveden is a change of use, the change is not such as to warrant major alterations to the appearance. The visitors to this particular hotel are seeking the atmosphere of an old country house, not significant change.

**Figure 7.5** (cont.)
(c) The original Library. (Copyright The National Trust.) (d) The Library converted to the dining room. (Copyright Derek Phillips)

(d)

(e)

(f)

**Figure 7.5** (cont.)
(e) Reconstruction of the dining room. (f) Picture lighting by fibre optics. (g) Dining room chandelier. (Copyright Derek Phillips.) (Architect Sir Charles Barry; lighting consultant Absolute Action Ltd for fibre optics)

(g)

## THE LANESBOROUGH HOTEL, LONDON

Converted from the original St George's Hospital, this is a far cry from Cliveden. Here the old hospital buildings, which had been unoccupied for twenty years, required extensive modification to make them suitable for a five-star luxury hotel.

The first St George's Hospital had been built on the site as early as 1734, but it was the later 1827 building by William Wilkins that has been modified. The new hotel interiors have been designed to give the impression of nineteenth-century opulence, and in no way reflect the appearance of the original hospital. This is an example of where the original building was a vehicle for its change of use, 'a valid well-daylit structure,' but where no attempt could have been made to express the original character of the hospital interiors. The exterior of the Wilkins facade has, however, been maintained.

The lighting designers concept comprised four elements:

1 That the apparent illumination should be seen to be provided by chandeliers, wall sconces, table- and floor-mounted light fittings, to achieve an 'authentic' atmosphere.
2 To provide acceptable operational light (enough light for the successful functioning of the hotel) by the integration of downlights recessed into ceilings, together with soft upward light often concealed within the decorative sources themselves.
3 To provide for special highlighting of important decorative elements within the interior by concealed miniature spotlighting to increase emphasis and modelling.
4 To achieve visual balance by the use of easily controlled sources (tungsten filament) which allow for the use of lighting control systems to vary the nature of the lighting to suit the time of day and hotel activities throughout the 24 hours of hotel operation.

**Figure 7.6**
The Lanesborough Hotel. (a) The Hall lit by pendants combined with table lamps. Drapes, flower arrangements and pictures create a subtle atmosphere. (b) The lift lobby. In addition to the floor-mounted torchère and pendant, low-voltage floor-recessed uplights are located at the pilasters.

(a)

(b)

Some examples of rooms will illustrate how the various concepts listed above have been employed.

In the Restaurant the pendant chandeliers, which in the past would have been candlelit, have a 'candle cup' incorporating a low-voltage lamp, the candle itself being a white glass tube, with flame top, glowing within the beam of light. Functional light is added by means of low-voltage recessed downlights, while key detail and visual drama provide enhancement of columns and window drapes. The visual balance reflects all the elements listed in the concepts outlined. In the conservatory, a highly stylized interior is reminiscent of the Royal Pavilion at Brighton. The same principles have been adopted, with the pendant chandeliers apparently providing the main light when in fact light is received from a series of other systems. In the great hall, or grand staircase, the same four elements are used and these can be clearly identified. The pendants provide the overall appearance of lighting the circulation space, downlights adding to the functional light, with the lighting of special elements such as flower arrangements, and with curtain lighting designed to balance the brightness of the daylighting seen through the windows, the whole controlled by a microprocessor 'scene set' system.

All public areas of the hotel can be analysed in this way, with the lift lobby and withdrawing room being excellent examples. The whole concept for the hotel, to provide nineteenth-century luxury in the twentieth century has clearly been achieved, and there can hardly be a more dramatic example of a change of use. With the exception of the exterior appearance, the original building has ceased to exist.

(c)

**Figure 7.6** (cont.)
(c) The Restaurant. The candelabra contains specially designed 'candles' to give an impression of real candles. Key detail and visual drama provide an enhanced atmosphere. (d) The Conservatory, with shades of the Royal Pavilion in Brighton. (e) The Withdrawing Room lit by a combination of lighting methods to achieve an authentic Edwardian atmosphere. (Architect Fitzroy Robinson and Ezra Atia (interiors); lighting consultant Lighting Design Partnership (Douglas Brennan); copyright Lighting Design Partnership)

(d)

(e)

# NORTH MIMMS PARK CONFERENCE CENTRE, HERTFORDSHIRE

The buildings with interiors by Ernest George, a pupil of Lutyens, originally used for stables to North Mimms Park Mansion, were converted for use by a brewery in 1988 as a conference centre for staff training, providing lecture rooms, dining hall, bar and reception lobby, together with a bedroom block. The concept for the centre was, where possible, to express the old timber structure but to provide comfortable conditions for the visitors, not unlike a country hotel. One of the factors governing the interiors was the need to integrate air-conditioning ducting in the lecture rooms. This could have been done by dropping a suspended ceiling below the cross-timbers and concealing the ducts above, but this would have denied the original concept to express the structure. The solution adopted by the architect was to carry the ducting between the roof trusses, allowing downlighting to be integrated in its design. If used alone this would have left the timber ceiling unlit, and a detail was developed for an upward light which reflected the design of the original corbels of the rafter supports. Flexibility was added by the capability of dimming the downlighting and omitting the uplight when special presentations were needed requiring minimum light levels.

An opposite effect is gained in the bar, where the original timber structure of the stables is expressed, the lighting being by pendant electric 'candlelight', backed up by low-voltage spots concealed among the beam structure at high level to give emphasis to the bar itself.

The dining room is not an historic building, having been added to the original structure to meet the requirements of the conference centre, but it was required to fit in with the general concept for the interiors. The lighting consists of pendant 'electric' candlelights suspended below the central glazed barrel vault with associated wall brackets, supplemented by low-voltage spots recessed in the surrounding lowered ceiling areas.

**Figure 7.7**
North Mimms Park Conference Centre. (a) The lecture room. Air conditioning is housed in a duct at roof level designed to incorporate downlights. The roof is uplit from side corbels. (b) The foyer lounge with downlights and modern wall lamps.

(b)

(a)

In the foyer serving the entrance and registration area there is a lowered ceiling with a staircase giving access to accommodation above. The space is lit with modern wall brackets, and additional recessed low-voltage downlights give emphasis to reception desk and sitting areas.

One further area which should be mentioned is not a part of the conference centre but is the old billiard room of the main mansion itself, where medieval paintings had been discovered. The room is now used as a boardroom, but a sophisticated system of lighting concealed around the perimeter of the new plaster ceiling has been installed to light up the restored paintings. The lighting is turned on rarely to conserve the paintings.

(c)

(d)

(e)

(f)

**Figure 7.7** (cont.)
(c) The Bar, all lit by spotlights concealed among the beam structure, together with a pendant candle holder. (d) The Dining Room, added to the original structure. (e) and (f) The Boardroom in the original mansion with light to medieval paintings. (Architect BGP; lighting consultant DPA; photographers Susan Hall with Derek Phillips; copyright Derek Phillips)

## THE BRIDGE CLUB, BURNHAM, BUCKINGHAMSHIRE

Another example of a change of use is the conversion of an old barn into a bridge club, with an associated bar and restaurant. Here the change may be radical but the original timber structure has been allowed to remain intact, despite the necessary addition of modern servicing, heating plumbing, etc., together with a new lighting system to meet the visual needs of the bridge players, without in any way reducing the impact of the original structure.

Externally the building has had extensive refurbishing to bring it up to present-day standards of weatherproofing and insulation. Being an old barn there was virtually no daylighting and this has been retained. It is the artificial lighting that has helped to create an interior which allows the timberwork (treated and cleaned up to show its mellow honey colour) to come alive and be enjoyed, at the same time as satisfying the building's primary function.

**Figure 7.8**
The bridge club, Burnham. (a) Exterior view of the barn as restored.

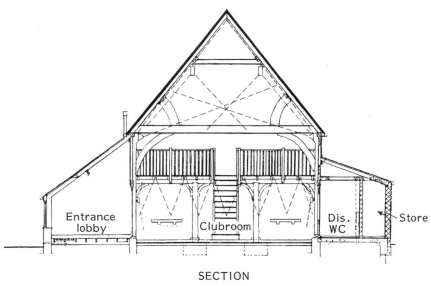

SECTION

**Figure 7.8**
(b) Section and plan of the barn to show the location of light fittings.

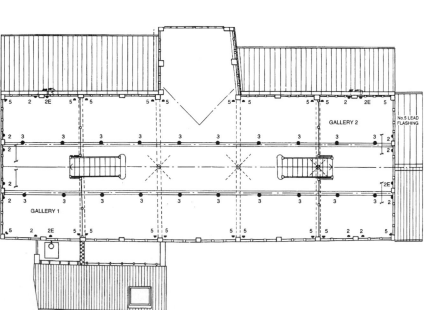

The method of lighting is in tune with the simple nature of the structure, comprising a series of pendant cylinders (basically 'black-painted cans' containing light sources) mounted on the parallel beams at high level, leaving the lower cross-beam structure clear. The latter act as visual baffles to help screen the lights from view, and it is from the level of these cross-beams that upward light is directed from the side-walls, across to the rafters of the roof structure above. The balconies at each end of the main 'playing' space, with the bar located below at one end, are new, but designed to fit well with the overall timber construction.

No attempt has been made to integrate the light fittings with the structure. All are face-mounted to separate them from the timbers, and where wall brackets are used these are placed in the centres of panels between the vertical timber wall supports. The little dining room has side windows for daylight and view and is provided with modern wall brackets, lighting both upward to the roof and sideways to the room. Downlighting is by means of low-voltage spots mounted at high level on the roof purlins directed towards the dining tables

All light sources are of the tungsten filament type, making dimming control simple. The lighting level in the dining room can be raised during the day, and lowered at night.

**Figure 7.8** (cont.)
(c) The Dining Room, an addition to the barn. (d) The main playing space seen from the first-floor balcony. (e) View from the club towards the bar. (Architect Hampton, Simpson and North; lighting consultant DPA; copyright Derek Phillips)

(e)

(c)

(d)

## NEW CONCORDIA WHARF, LONDON

Close to Tower Bridge on the Thames, the New Concordia Wharf building was constructed in 1885 at the same time as the bridge itself, named after a town called Concordia near Kansas City in Missouri. The warehouse was sold in 1934 to the Butlers Wharf Co. and was used for some years for the storage of tea and other goods.

By 1980 the warehouse use had become redundant, and following restoration and conversion New Concordia was marketed as 'shells' of varying size for domestic development, studios and offices. The overall plan shows the subdivision where some of the spaces have the advantage of bilateral daylighting while some, such as the one illustrated in Figure 7.9(a), have daylight from one side only. For this reason some compromise had to be adopted, first to maximize the daylight available by placing the main living space along the available outside wall, and then to use artificial lighting for internal areas, such as entrance hall and dining room. The kitchen, bedroom and little library gain from daylight borrowed from the living room, but rely on artificial light both during the day and at night.

The appearance of the spaces is influenced by the original warehouse construction, expressed by the exposed brickwork walls and timber floors, together with cast iron columns. The artificial lighting makes no attempt to conceal the light sources, but chooses those which can be placed between the timber floor beams for directing at walls for general light, with low-level portable lamps. All lights are designed to eliminate glare from normal viewing angles.

**Figure 7.9**
New Concordia Wharf. (a) Plans of the warehouse showing both the overall subdivision of one level and the detailed planning of a flat with daylight from one side only.

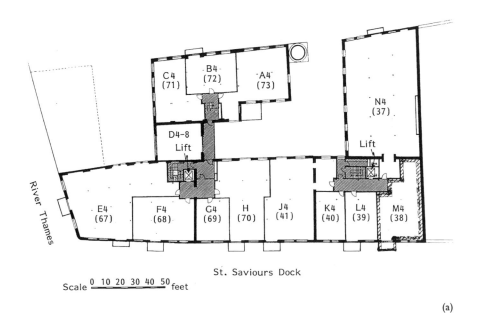

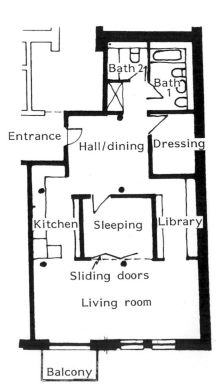

(a)

(b)

(c)

(d)

(e)

**Figure 7.9** (cont.)
(b) Exterior as seen from the Thames. (c) Interior of living room in daylight supplemented by artificial light on a dull day. (d) The living room at night. (e) The kitchen, gaining borrowed daylight from the living room as well as requiring artificial light when in use. (Architect Pollard Thomas Edwards for the reconstruction; copyright Derek Phillips)

## BILLINGSGATE FISH MARKET, LONDON

The original Victorian building, designed by Sir Horace Jones and housing the famous London fish market, was converted into a city dealing room in 1985. Due to the difficulties of the London financial markets at that time, it remained empty until it was taken over by the NatWest Bank in 1993 as a duplicate computer facility.

The planning to the building necessitated the installation of additional space, and this was overcome by the adoption of a suspended floor. The change of use from a daylit market required great ingenuity since the direct daylight received through the overhead glazing was unsuitable for office space serving computer VDUs.

Special prismatic plates developed in Germany were used to modify the direction of the natural light to allow the basic daylight strategy for the building to continue, thus perpetuating the original overall visual quality of the space. The prismatic plates installed within new double-glazed units which replaced the original glazing allow the entry of natural light, with all its psychological and economic advantages, while reflecting out direct sunlight. In the words of the architect 'this produces a calm even light, devoid of drama, but which reacts to the dynamics of daylight'. The fluorescent source, consisting of continuous lines of specially designed reflector fittings, relate to the overhead beam structure. These provide the downward light, controlled to eliminate glare, but also allow some upward light to the roof structure. The whole system is controlled by photocells to ensure that when the daylight drops below an adequate working level the artificial light is brought in to supplement it.

The Haddock Gallery at high level benefits from overhead roof glazing in which the same system of prismatic panels controls sunlight. Natural light alone is used during the day and a system of indirect light concealed above the arches provides adequate light for an office using VDU screens.

The redevelopment of Billingsgate is a good example of how many of our daylight-oriented historic buildings can be given useful new life using state-of-the-art lighting technology, without in any way destroying their intrinsic spatial qualities.

**Figure 7.10**
Billingsgate. (a) View of exterior from across the Thames after conversion. (Photographer Eamonn O'Mahony; copyright Lord Rogers Partnership.)

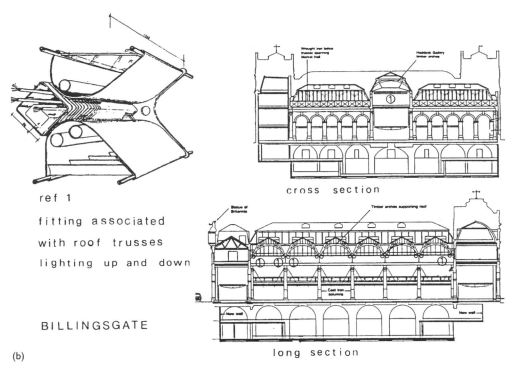

ref 1

fitting associated

with roof trusses

lighting up and down

BILLINGSGATE

(b)

cross section

long section

(c)

(d)

(e)

**Figure 7.10** (cont.)

(b) Section through the building indicating the new fluorescent lighting installation. (Copyright Lord Rogers partnership.) (c) Interior of main Computer Hall ready for occupation by the NatWest Bank. (Copyright Derek Phillips.) (d) The Haddock Gallery occupied. A system of uplighting, using the 250 watt metal halide lamp, is placed at high level above the arches, providing indirect artificial light at night. (e) The prismatic panels controlling daylight (Copyright Derek Phillips.) (Architect Lord Rogers Partnership (for the conversion); lighting consultant Lighting Design Partnership)

# 8  Exteriors

The exterior lighting of historic buildings (generally thought of as floodlighting) is often more successful than with modern structures, due to its decoration, detail and modelling, and the greater proportion of solid to void of the facade. Indeed, the interior lighting of many modern buildings, seen as a counterpoint with the more limited solid areas, can render external floodlighting superfluous.

Floodlighting is a comparatively modern concept, and the original building would not have been conceived with exterior lighting in mind. There is therefore no traditional way to go about it, other than the methods that have been adopted over the past hundred years, which have established their own momentum. Important aspects to bear in mind when establishing the brief are as follows: the relationship with the site and the possible view positions, function, modelling, colour and glare.

## THE BRIEF

A study of the daylight appearance of the building is an important part of the brief to the lighting designer. The daylight appearance of the exterior will help an understanding of what should be, or should not be, emphasized at night.

A careful analysis of the building, its surroundings and its social implications is essential before decisions are made as to how the lighting both internally and externally should be approached. This is not to suggest that the night-time view should imitate the appearance of the building during the day, rather the reverse. In analysing what is most appropriate, the significant features of the building should be emphasized ... if it is round, its rotundity; if it is tall, its height; if it is asymmetrical, its form, etc. To a degree, the lighting designer can correct any faults the building may have by not emphasizing certain elements at the expense of others, and for this reason the building may appear more attractive by night than during the day. Its appearance will in any case be different, but the building itself should always appear to be the same building after dark as during the day.

The views of the building whether from close up or at a distance must be studied, as the available viewing angles have a direct influence on the possible and desirable location of equipment when assessing its accessibility and the problem of glare from the light sources to be used. This is often one of the determinants of the design, as to what can and should be achieved.

The social implications of the night-time lighting of historic buildings, seen in their context, must be considered. For example, a church in a country setting, where it is in no competition with its surroundings, requires only moderate lighting but its position in the village, perhaps providing a vista seen from an approach road, gives it added significance. In towns the approach to the building within its setting may provide a vista which leads the eye towards it. So the setting of the building has great social significance, something which needs to be addressed when establishing the brief to the lighting designer.

Historic buildings often offer a wealth of decoration, the modelling of which can be emphasized and highlighted. The brief must establish the aims to be met by the exterior lighting, whether for security, for attraction, or for specific views. The brief should establish the environment into which the building will be set (rural, urban, etc.) and the appropriate colour to harmonize or to contrast with its surroundings.

One of the problems that the lighting designer will have is to convey to the building owner how the building will look at night. Computer studies can assist here but these are at present very expensive and time consuming, whereas one of the early methods used was by the creation of a perspective drawing illustrating the scheme as it might appear by night.

An example of how this was done in the 1970s, was for the winning entry in a competition for the floodlighting of Westminster Abbey. Figure 8.2 (a) is an impression by the designer of how the Henry VII Chapel

**Figure 8.1**
Proposal for relighting (a) Edinburgh

(a)

the building should express both integrity and unity. Different surfacing materials require different sources – warm sources for warm stonework or brickwork, cool sources for white stone or concrete. It is important that this be seen as the normal approach, but this does not preclude the abnormal in certain circumstances. There may be a case for exceptions to be made, as, for example, for festive lighting. However, this should be done for sound and well-thought-out reasons, and not be accidental.

Caution must be exercised over the use of colour, as it can go so badly wrong, and one of the ground rules is that colour should be limited where the fabric of the building is permanent and related to the surface colour of the work. For ephemeral effects, such as the lighting of parks and water at certain seasons of the year, or for festive occasions, like 'fireworks' a bolder use of colour may be appropriate. A useful rule is 'when in doubt leave it out', but this should never preclude the excitement that can be displayed for some special occasion by throwing patterns of light onto buildings at night.

(a)

(b)

**Figure 8.3**
(a) The Statue of Liberty, New York. Fireworks for the bicentenary celebrations and completion of the new floodlighting. (Lighting consultant H. Brandston Lighting Design; copyright General Electric USA.) (b) Proposals for the Son et Lumière for the Palazzo Reale, Caserta, Italy. Images of Leonardo da Vinci's 'Last Supper' projected onto the face of a building. (Lighting consultant and copyright Jonathan Speirs Associates; computer simulations by Photoshop)

## GLARE

Glare should always be avoided, and if, as in some city sites, it is impossible to place light sources in situations where this cannot be done then they should be omitted. Glare raises the adaptation level of those viewing the building to a point where it is no longer possible to enjoy the nightscape. When there is an absence of glare, the environmental level of light can be lowered, a further example of 'a little light going a long way'.

The presence of glare means uneconomic solutions where light is wasted. When light is wasted, it generally means that there is 'light pollution' or light which is unwanted and which affects others adversely. Another name for this is 'light trespass'.

The elimination of glare is really the key, since so many famous buildings have been floodlit in such a way that they exclude glare from some viewpoints but not from others. What is one person's delightful vista is another's visual pollution. There is no excuse for this, and yet some of our most famous buildings and bridges suffer in this way. Figure 8.4 shows historic buildings that have been lit at night.

**Figure 8.4**
Floodlighting examples. (a) Notre Dame Cathedral, Paris. (Copyright *International Lighting Review*.) (b) Plaza de Toros, Valencia. (Copyright Iguzzini Lighting.) (c) Albert Bridge, London. (Copyright *International Lighting Review*)

(a)

(b)

(c)

## THE GRAND' PLACE, BRUSSELS

The Grand' Place (main square) in Brussels was an early example of the floodlighting of a major building, the sixteenth-century Town Hall, at one side of the square. Lighting equipment has been integrated with the structure at the higher levels and on the roof to light up the tower. These sources provide modelling to the intricacy of the decoration of the facade. Where it is impossible to light the lower storey without glare from equipment mounted at pavement level special street-lighting columns have been adapted to solve the problem. These provide normal street

(b)

(a)

**Figure 8.5**
Grand' Place, Brussels. (a) General view of the Town Hall. (b) Detail of the pedestrian street lamp incorporating floodlighting source. (Lighting consultant Philips Lighting; copyright *International Lighting Review*)

light for pedestrians, but, in addition, floodlights are integrated within the tops of the columns to light up the lower storey of the building. It is surprising that this type of solution has not been adopted more often, since the presence of pedestrian street lamps is an acceptable part of the urban landscape during the day while large unsightly floodlighting equipment mounted at low level is not.

## TREVI FOUNTAIN, ROME

Perhaps the most famous of the fountains in Rome is the 'Fontana di Trevi' dating from 1878. This is a perfect example of Baroque art, lit by a method not dissimilar to the Grand' Place in that 1000 watt floods are concealed in the nineteenth-century lanterns which border the flight of steps leading down from the square to the fountain, thus using pedestrian street furniture to provide concealment for the floodlighting.

The light from these lanterns gives the overall environmental light to the statue of Neptune and the sea gods, with the tall decorative wall and columns behind the water. The fountain basins are the location for lighting equipment concealed below the water, lighting the rocky front and the water jets of the fountains themselves, together with the lively surface of the water itself. The limited viewing positions of the fountain make the control of spill light and glare relatively simple.

**Figure 8.6**
The Trevi Fountain in Rome (a) by day and (b) floodlit by night. (Lighting consultant Philips Lighting; copyright *International Lighting Review*)

(a)

(b)

## COUR CARRÉE, THE LOUVRE, PARIS

The night-time lighting of the 'Square Court' at the Louvre uses entirely new techniques, both in the manner of its design and in its implementation. The design technique chosen uses computer-generated three-dimensional virtual images to simulate the night-time lighting of the building, before any lighting trials have been carried out. While most floodlighting of historic buildings directs light upwards from below, inverting the daytime appearance where the daylight comes from

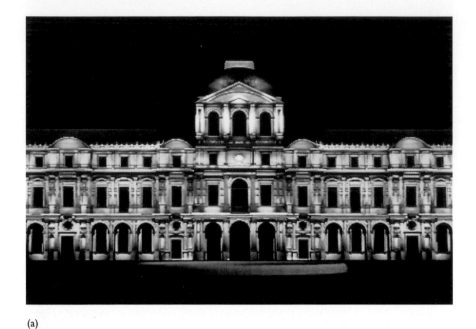

(a)

**Figure 8.7**
Cour Carrée, the Louvre. (a) Computer simulation of the exterior lighting effect, used as the basis for developing equipment to meet the design criteria. (Copyright CRAI, Nancy.) (b) The completed floodlit facade. (Copyright *International Lighting Review.*)

above, it was possible by this technique to simulate light from above to allow a visualization of the building lit by lighting angles similar to sunlight.

Having agreed that this appearance was what the client wished, the next stage was to design special lighting equipment to achieve the lit effect, shown by the computer simulation. A special range of lighting equipment was developed for this purpose, consisting of a miniature linear reflector system which could be attached to the under-surface of cornices to cast the light from miniature linear low-voltage small-wattage lamps placed end to end, lighting downwards.

It is interesting to see how closely the final appearance conforms to the computer simulation. The lamps planned are said to have a life of 20 000 hours, so the problem of changing lamps is reduced. It should be emphasized that such a lighting system has been designed specifically for

(b)

this building, relying on placement below the available projecting cornices and is unsuitable for universal use.

The fitting developed specially for this project can be seen when looking up at the building from close to but not at normal viewing angles. The design method, however, whereby a building is created in the computer and then different lighting patterns applied, must have much to recommend it. As with the artist's impressions illustrated earlier, it can show the building owner how it will look, but at present the use of computers for floodlighting design is expensive, and this does not necessarily solve the lighting method or the design and location of the hardware, the means to create the desired effect.

**Figure 8.7** (cont.)
(c) Section of the linear fitting developed. (Copyright *International Lighting Review.*) (d) Detail of the linear fitting. (Copyright Crescent Lighting.) (e) View looking up towards the lighting equipment. This is not a view normally available to the public.(Copyright *International Lighting Review.*) (Lighting consultant Electricité de France (Roland Jusot and Olivier Wattiau))

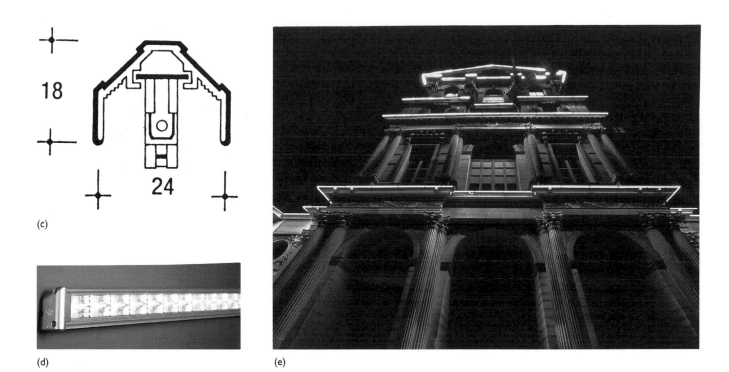

(c)

(d)

(e)

## ROMAN BATHS, BATH

The juxtaposition of masonry with water, fire and light is a magic combination, and the Great Bath in the Roman remains at Bath is no exception. The thermal springs which gave rise to the famous baths were discovered in the first century by the Romans, who used their mineral properties for 400 years. After they left, it took 1200 years before the British aristocracy rediscovered their healing properties in the eighteenth century.

The relighting of the historic complex of the Bath Museum, Pump Room and archaeological remains in 1980 brought the whole complex to life. This was an exercise where the designs involved both interior and exterior lighting, with the difficult relationship which can exist between the two, when a person used to bright sunlight outside during the day must adapt to the much darker interior. However, our principal concern is the night-time lighting of the exterior of the courtyard of the Great

Bath. The light derives from light sources placed around the covered arcade which surrounds the bath, leaving the central area of water unlit. But to add to the magic, flame sources are placed around the water one to each column, using the 'fire' from gaslight.

**Figure 8.8**
The Great Bath in the Roman remains of Bath at night. (Lighting consultant Philips Lighting (J. W. Carlton and T. D. J. Hughes); copyright *International Lighting Review*)

## LEEDS CASTLE, KENT

This is one of the most beautifully sited castles in Britain, rebuilt in the fourteenth century by Edward I and enlarged by Henry VIII. It is now used for government conferences, and is open to the public at other times. The castle is surrounded by an artificial lake, fed from the river Len, and when the surface of the water is still, the views across the lake benefit from reflections of the castle during the day and now at night.

(a)

**Figure 8.9**
Leeds Castle. (a) The castle by day, seen across the lake.

The castle can be viewed from all sides of the lake, and since the brief called for the three main sides of the castle and surrounding walls to be illuminated, a degree of compromise was necessary. It would be impossible to prevent views of some floodlights on one side of the lake being seen from the other. However, the way in which the floods have been located and directed minimizes this from all but a few positions. In addition, the direction of the floods has been designed to give modelling to the facade.

Great care has been taken to conceal the floods below eye level, in the surrounding banks of the lake, and as the rural setting presents no competition, the level of light can be low to achieve a dramatic effect. A very economical electrical layout has been designed, only five 1 kilowatt floods being used to achieve the whole effect, spaced around the edge of the lake. The colour of the floods has been carefully chosen to enhance the Kent ragstone of which the castle has been constructed.

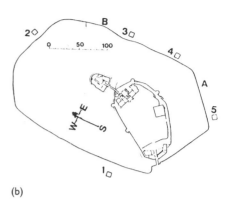

(b)

(c)

**Figure 8.9** (cont.)
(b) Plan of the castle, showing the location of the floodlights. (c) Floodlighting from the south-east. (d) Floodlighting from the north-east. (e) Detail of the floodlight embrasure set into the verge of the lake. (Photographer Alan Turner; lighting consultant and copyright Thorn Lighting)

(d)

(e)

## PRINCE EDWARD THEATRE, LONDON

The interior of the theatre has been discussed in Chapter 6, but the design for the exterior lighting is an example of the subtle use of light on the face of a building, where all the light sources are incorporated within the structure of the facade itself. The lighting design was facilitated by the renovation of the entrance canopy to the theatre, which allowed a number of lighting systems to be incorporated within the canopy. Deeply recessed downlights were integrated into the under-surface to emphasize and provide a welcome to the entrance area. Upward light sources were provided to light the underside of the recessed elevation and narrow-beam spots light the columns. The whole face of the canopy is also lit, associated with the lettering to the advertising sign for 'Prince Edward Theatre'. The interior lighting of the entrance hall 'arches' seen from the outside adds further emphasis to the facade at low level.

Above the entrance canopy, wall-mounted torchères are placed, lighting up to the brickwork, and the attic storey has upward light concealed above the projecting cornice to increase the apparent height of the facade. The whole lighting design produces a sophisticated exterior appearance, which greatly adds to the attraction of the theatre.

**Figure 8.10**
Prince Edward Theatre. (a) General view of the exterior at night. (b) Detail view of the entrance canopy. (Architect RHWL for the restoration; lighting consultant and copyright Lighting Design Partnership)

(a)

(b)

# PIRELLI GARDEN, VICTORIA AND ALBERT MUSEUM, LONDON

The brief to the lighting designers for the Pirelli Garden was a response to the need for the museum to become more financially self-sufficient. This was to extend the use of the garden and to make it suitable for evening functions, enabling it to be let out for client entertainment and other promotional activities.

The final night-time lighting scheme is an example of the build-up of different layers of lighting to achieve a unified whole. One of the chief aims of the designers was to ensure that the equipment used should be unobtrusive, both by day and by night, and that it should be installed in such a manner that no damage was done to the delicate structure of the building. Figures 8.11 (a)–(d) illustrate the way in which the lighting was built up. First, the door pillars and windows are uplit by narrow beam spotlights and the tiled arches lit by concealed linear fluorescents. Together these produce a theatrical glow which appears to radiate from the interior of the building. The pediment, roof statues, and areas of the building's facade, are floodlit from adjacent buildings, using powerful narrow-beam floods. These allow the light to be precisely focused and avoid the danger of glare to those in the garden. Finally, the trees are lit from buried floodlights around the bases of the trees, and the fountain lit from the roof of the opposite building by narrow-beam floods. The fountain can be covered over as a podium for presentations.

**Figure 8.11**
Pirelli Garden. (a) A daylight view of the main building.

(a)

The light sources used are predominantly filament types of tungsten or tungsten halogen, producing a soft warm light, sympathetic to the colour of the brickwork. These light sources assist the different elements to be set up and balanced against each other by means of an elaborate control system, allowing a variety of dramatic effects, achieving the purpose of the brief to provide a valuable new resource for the museum.

**Figure 8.11** (cont.) (b) The lighting of window embrasures. (c) Frontal light to the facade. (d) Combined lighting scheme with the tree lighting. (Lighting consultant and copyright Lighting Design Partnership)

(b)

(c)

(d)

## THE STATUE OF LIBERTY, NEW YORK

For most people arriving in New York, the Statue of Liberty was the symbol of hope in the New World, and it is this symbolic aspect which inspired the lighting design. Built in 1886 to the designs of the sculptor Frédéric-Auguste Bartholdi, the statue has been lit and relit several times, the latest being in 1986, for its centenary.

To understand the scale of the project it is necessary to realize the size of the statue needed to dominate the harbour, and this is best understood by looking at a section through the statue, with its accompanying base structure. The statue is immense, as can be seen from the many levels, and was for some years the tallest structure in New York, rising to 100 metres. To any lighting designer it would be a daunting project, but by an understanding of its history, and an intimate knowledge of its form derived from close visual studies, the designer has produced a masterpiece.

The relighting was associated with the restoration of the statue and this in itself was a fascinating story of a technological breakthrough to overcome the ravages of time, but structural technology was not the only breakthrough. Lamp and lighting technology was developed specifically for the project.

The lamp industry in the USA developed a new metal halide lamp to give lighting designers the colour, intensity and directional qualities they wanted, allowing the positioning of the lamps to be set well away from the base (100 metres) and to give precise aiming to bring out the statue's form. The lamp developed gave the right reflected colour from the green of the copper plate from which the statue is made, and allows alternative warmer and cooler effects.

Pits were constructed at a distance from the statue, concealing the floods, special angled grills being placed over them to control the light. The light is then directed in such a manner as to increase its intensity towards the top of the statue, a most important aspect of the design when the statue is seen from across the harbour. There is no awareness of where the light is coming from, it is as though 'the sun has not set'.

It is of interest to see that one of Bartholdi's initial ideas had been to have a halo of light beams emanating from the crown of the statue. This had not been technically possible in 1886, but the consultant thought that it would have been possible in 1986, and produced a sketch to illustrate the idea. However, this was dropped for a more orthodox appearance.

The flaming torch at the top, held in the hand of Liberty, had originally been a solid form but it was later perforated with windows letting out light. This had not proved a success, and a new torch was developed clad in gold leaf. Instead of the light shining through, light was reflected from lamps directed from the outside concealed in the base of the torch, giving it a much greater impact.

Some facts on the issue of energy for relighting the statue are relevant. The lighting scheme for the statue in 1916 used 61.5 kilowatts of energy and cost $20 500 to run for a year. In 1976 the scheme used 68.6 kilowatts of energy, and cost $22 800 000 to run for a year. The 1986 scheme uses only 17.9 kilowatts of energy and costs $6 000 to run for a year. This is a tribute to the developments in lamp technology, and a useful point to bear in mind when the floodlighting of such monuments is discussed.

To conclude, the lighting of the Statue of Liberty exhibits all the important characteristics set out for good exterior lighting. A brief was

**Figure 8.12**
Statue of Liberty, New York. (a) Section through the statue to indicate its size. (b) Plan indicating the site.

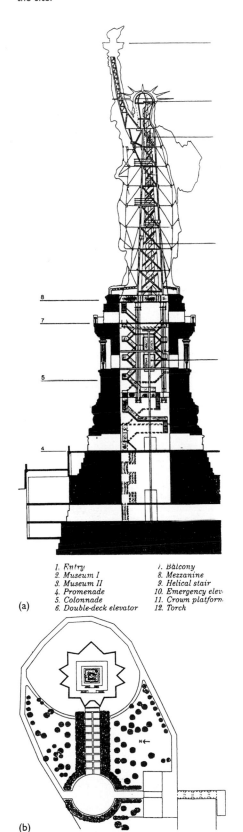

1. Entry
2. Museum I
3. Museum II
4. Promenade
5. Colonnade
6. Double-deck elevator
7. Balcony
8. Mezzanine
9. Helical stair
10. Emergency elev.
11. Crown platform
12. Torch

(a)

(b)

(c)

(d)

worked out between the City of New York, the architects and the lighting designer to achieve an effect derived from its daytime appearance, when seen from long distances across the harbour, as well as to visitors to the museum housed in the base of the statue. This was a functional scheme, capable of being installed and maintained to provide long service, from fittings housed in special enclosures where they are protected, taking into account the needs of energy management. The modelling of the statue, the folds of its garments and the shape and character of its head, succeed in satisfying the brief, with additional points of interest in the lit crown and the gold flame. The new light sources developed especially for the project have been related to the colour of the materials of the statue to bring out the patina of the copper plate. There is a complete absence of glare, with all light sources being concealed.

It is hoped that this chapter has emphasized the need for the same understanding of the nature and function of historic buildings, when seen from outside after dark, as is required for their interiors.

**Figure 8.12 (cont.)**
(c) Original concepts as sketched by the lighting consultants influenced by the vision of Bartholdi.
(d) View of the statue at night, with the emphasis on the folds of the garment clearly seen. (Lighting consultant Howard Brandston)

# Appendix

## CONSERVATION

It is not the intention in this brief appendix to do more than emphasize the importance of the need to consider the long-term effects of both daylight and artificial light on the objects within an historic building. In the words of Bernard Fielden in his excellent book on the subject, 'conservation demands wise management of resources, sound judgement and a clear sense of proportion . . . it demands the desire and dedication to ensure that our cultural heritage is preserved'.

The recognition of the need for conservation is of comparatively recent date. It was not until the twentieth century that people became fully aware of the deterioration caused to the interior elements of buildings by 'light', both by day and by night. Although the effects of 'ageing' must have been experienced, it would not have been understood that this was caused by the introduction of ultraviolet light from the natural rays of the sun through the window. After dark, the amount of light from the artificial sources of the day would not have been of sufficient brightness to cause deterioration. A more likely cause would have been due to the products of combustion, from oil, candles, and, later, gas lamps, and there are instances of complaints of fine interiors being spoilt by the smoke of oil lamps as early as the eighteenth century.

There is a clear relationship between the level of exposure to light and the extent of deterioration occurring to materials. This deterioration is caused by two factors, thermal and photochemical. The thermal effects are caused by the infrared content of the light source, whether daylight (sunlight) or artificial, which causes changes in relative humidity in the region of the lit object, the heat from which over time will result in the warping or splitting of organic materials. The photochemical effects are caused by the ultraviolet content of the light, chiefly from daylight, which affects the colour of dyes and pigments. The light from tungsten filament sources is considered not to pose a threat, as the ultraviolet content is minimal.

Before any relighting of an historic building is undertaken, an assessment should be made of the materials used in the interior, and particularly of those most sensitive to light, and to relate these to known safe limits of light. The subject of conservation of materials is covered in the CIBSE Lighting Guide, *Museums and Art Galleries*, where the following guidance is given for the illuminance of the three main categories of object:

# Glossary

This glossary contains words which the author considers might be subject to a different interpretation, or have a specific technical meaning, such as those used in architecture or lighting design.

## LIGHT SOURCES

**Candles** Candles are made by moulding wax or other flammable material around a wick, which sustains a flame to give light. The earliest candles were produced by dipping a rush taper in melted fat, but later developments used moulds with tallow. Subsequently more refined materials such as wax or spermaceti were used so that the candles today are clean, do not gutter, and provide light of a particular quality suitable for social occasions, such as dining. The demise of the candle as a light source has long been predicted by the lighting industry, happily without success.

**Daylight** The light received from the sun and the sky, which varies throughout the day, as modified by the weather. The daylight within a building is modified by both its surroundings and the nature of the apertures (windows) through which it enters, a proportion only of the daylight outside being available inside. Daylight has ·well-known characteristics of direction and colour which determine what is understood as 'natural light'. There are no absolutes in terms of measurement, the daylight inside being a 'factor' of what is available, which is itself variable. Those environmental factors of daylight which contribute to quality are well covered in the text.

**Electric light** All modern artificial light uses electricity as its primary source, but from the arc lamps developed in the early nineteenth century to the sophisticated light sources today there is a world of difference. A brief description of electric light sources follows.

**Arc lights** Passing an electric current between two electrodes causes a spark or arc, giving off light. This early method of generating light from electricity was developed by Sir Humphry Davy in 1809. It was cumbersome and tended to be used in lighthouses and for sports arenas in the nineteenth century. Carbon arcs, as they were called, had limited use in some interiors at the turn of the century before the introduction of incandescent or filament lamps.

**Cold cathode lighting**   Similar in many respects to fluorescent lighting, the cold cathode source operates at high voltage, making certain safety requirements necessary. Whereas there are difficulties in dimming normal fluorescent, cold cathode is easily dimmed. Another of its advantages is that it can be formed to take up curves and other shapes to fit into architectural detail, and a variety of colours is available. Where used in a concealed cornice, continuous gap-free lines of light can be formed, but it is not widely manufactured.

**Fluorescent lamps**   The fluorescent lamps developed after the Second World War suffered from poor colour rendering. They were long and bulky, and required control gear, limiting the opportunities for dimming. Many historic buildings were converted at this time to fluorescent lighting due to the economics of long lamp life and greater efficiency; but architecturally fluorescent fittings were difficult to handle, and much fluorescent lighting has now been removed and replaced with other forms of light, more amenable to the architecture of historic buildings. This should not preclude the use of fluorescent lamps where their characteristics are appropriate in those buildings for which a change of use is planned (for example, in a conversion to modern offices). Modern fluorescent lamps are available in a variety of lengths, diameters and colours. Colour rendering has been improved, though still not ideal, and small 'compact' lamps are available, which make a useful contribution to the designer's palette. Their greatest attribute is their long life (up to 8000 hours) and their efficiency (from 60 to 100 lumens per watt, bearing in mind the efficiency of tungsten of only 12–14).

**ISL lamps**   The 'internally silvered reflector' lamp is a less expensive alternative to the PAR, and available in a variety of sizes and wattages. They are unsuitable for exterior use.

**Low-voltage lamps**   Normal lamps operate on mains voltage (230–240 volts) but low-voltage (LV) lamps operating at 12–24 volts are available in filament ranges. Their chief advantage is in their small size, allowing miniaturization of fittings and reflector systems. Lamps are available as tungsten halogen types, either with or without an integral reflector. Due to the ease with which they may be concealed in small spaces, these lamps have a special use in the relighting of historic buildings. It must be remembered that they require transformers to convert mains to low voltage, and these need to be hidden close by where they can be maintained. (Note: In the USA the mains voltage is only 110 volts, so that lamps and fittings are not compatible without modification.)

**PAR lamps**   The term means 'pressed glass aluminized reflector' and refers to those sealed-beam lamps where a parabolic silvered reflector provides accurate beams of light. The most common kinds are the PAR 38 lamp (varying from 60 to 120 watts) or the PAR 56 300-watt lamp. As the beams are sealed they are less subject to dust, and lamps of different beam widths are available. They are useful lamps where access for cleaning is difficult. They may be used externally.

**Tungsten filament lamps**   Known generally as light bulbs, the tungsten filament lamp is the most common of all electric sources. It is the cheapest and the least efficient. When originally developed by Edison in the USA and Swan in the UK, it was seen as a magical source, despite its early life of as little as 150 hours and low efficiency. The life of the lamp now is 1000–2500 hours with an efficiency of approximately 12–14 lumens to the